Ferns of Hawai'i

Ferns of Hawai'i

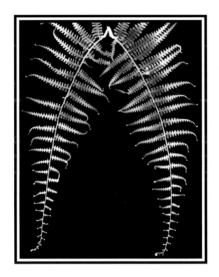

KATHY VALIER

A *Kolowalu* Book
UNIVERSITY OF HAWAI'I PRESS
HONOLULU

© 1995 Kathy Valier
All rights reserved
Printed in Singapore

95 96 97 98 99 00   5 4 3 2 1

Figures © 1995 Chris Faye

Library of Congress Cataloging-in-Publication Data

Valier, Kathy, 1953–
Ferns of Hawai'i / Kathy Valier.
p.   cm.
"A Kolowalu book."
Includes bibliographical references (p. 00) and index.
ISBN 0-8248-1640-4
1. Ferns—Hawaii—Identification.   I. Title.
QK532.H4V35   1995
587'.3'0996—dc
94-38833
CIP

University of Hawai'i Press books are printed on acid-free paper and meet the guidelines for permanence and durability of the Council on Library Resources

Designed by Paula Newcomb

CONTENTS

Preface / vii

INTRODUCTION

Ferns in Hawai'i / 3
How Ferns Differ from Other Plants / 6
Fern Life-styles / 6
Designs in Nature / 8
Spores Instead of Flowers and Fruit / 10
What Distinguishes Different Ferns / 10
What's in a Name? / 12
Where to Find Ferns / 15

FERN ALLIES

Club Mosses / 19
Whisk Ferns / 21
Spikemosses / 22

COMMON FERNS

Adder's Tongue / 25
False Staghorn Ferns / 26
Finger Ferns / 29
Polypody Ferns / 33
Cliff Brake Ferns / 37
Pteris Ferns / 39
Maidenhair Ferns / 41
Silver Ferns / 45
Filmy Ferns / 47
Tree Ferns / 51
Hay-scented Ferns / 53

Bracken Ferns / 54
Lace Ferns / 56
Maiden Ferns / 57
Spleenwort Ferns / 60
Lady Ferns / 63
Wood Ferns / 68
Stag's Tongue Ferns / 71
Sword Ferns / 74
Blechnum Ferns / 76
Water Fern / 80

Glossary / 83

Bibliography / 85

Index / 87

Color illustrations follow pages 8, 24, 40, and 56

PREFACE

Ferns form the verdant backdrop for the colorful trees and shrubs of Hawai'i. Although they don't have showy flowers or edible fruit, the varied designs of ferns are beautiful and worth taking the time to stop and appreciate.

For thirty-six years, I hiked through the myriad ferns around me, mindlessly relegating the 200 wild ferns and fern allies in Hawai'i to the general category of "ferns." Only one tangled scrambling fern, *uluhe*, stood out from the rest, because it chokes trails and scratched my legs. The need to identify one species, *palapalai*, for a project made me realize how varied these subtle plants are and how little I knew about them. I gathered the material in this book to educate myself on the wealth of ferns in Hawai'i.

Hawai'i is ideal for studying ferns. They grow year-round in the tropical climate, and the diverse settings on each island allow a wide variety of ferns to grow in a relatively small area. The famous botanist David Douglas, for whom the Douglas fir is named, described the numerous ferns he discovered on a trip up Mauna Kea in 1834:

> The path was in a dreadful state, numerous rivulets overflowed it in many places, and, rising above the banks, rushed in foam through the deep glens, the necessity for crossing which impeded my progress in no slight degree . . . still, violent as was the rain and slippery and dangerous the path, I gathered a truly splendid collection of ferns, of nearly 50 species, with a few other plants, and some seeds . . . and in spite of all the disadvantages and fatigue that I underwent, still the magnificence of the scenery commanded my frequent attention, and I repeatedly sate [sic] down in the course of the day, under some huge spreading Tree-fern, which more resembled an individual of Pine [palm?] than the Fern tribe, and contemplated with delight the endless variety of form and structure that adorned the objects around me. (Wilson 1922)

Ferns grow in delightful landscapes. An English fern-lover, Frances Theodora Parsons, wrote in 1899:

> To me the greatest charm the ferns possess is that of their surroundings. No other plants know so well how to choose their haunts. If you wish to know ferns you must follow them, to Nature's most sacred retreats. In remote, tangled swamps, overhanging the swift, noiseless brook in the heart of the forest, close to the rush of the foaming waterfall, in the depths of some dark ravine, or perhaps high up on mountain-ledges, where the air is purer and the world wider and life more beautiful than we had fancied, these wild, graceful things are most at home. (P. 13)

Studying ferns in the wild can give you hours of pleasure and heighten your awareness of nature's intricate designs. This book aims to enhance your appreciation of ferns by focusing on their finer aspects. You need not be a botanist to enjoy the ferns of Hawai'i. A glossary is provided. I hope that this information will bring you to a greater appreciation of these delicate works of nature and enrich your explorations of Hawai'i.

I am indebted to Tim Flynn and Charles Lamoureux, who initiated my interest in ferns and who, along with Dave Lorence, Bob Hobdy, and Kay Kepler, encouraged me to write this guide. Warren H. Wagner, Molly Sommers, Joel Lau, Dan Palmer, and Roselle Bailey reviewed the work and contributed their knowledge to the text.

Many special companions hiked to fern haunts with me: Debbie Ward, Bob Nishek, A. J. Metzgar, Brian Schaefer, Norman Mitchell, Suzanne Marinelli, Peter Novis, and my mother, Marjorie.

Introduction

Hawai'i, with its wet, tropical climate and diversity of landscapes, is a showcase of ferns. In few other places do such a variety of ferns grow in such a small geographical area. Approximately 200 species of ferns grow in the 6,400 square miles (16,576 sq km) of Hawai'i. In comparison, New Zealand, which abounds with ferns, has as many species, but they are spread over sixteen times as much land. This book includes sixty-one of the most common ferns and fern allies growing wild in Hawai'i.

Ferns in this book are common on all the Islands, except where specifically stated in the description. A variety of ferns was chosen, to represent the many families of ferns found in Hawai'i and to show the wide range of forms Hawaiian ferns can take, from the tiny water fern, *Azolla*, to the wiry masses of scrambling false staghorn ferns (*uluhe*). Most of the ferns used by Hawaiians are included. Ferns are listed in order from primitive to more highly evolved.

FERNS IN HAWAI'I

As the Hawaiian Islands formed, ferns were often probably the first vascular plants to colonize fresh lava flows, crossing wide stretches of ocean as wind-blown spores. Self-introduced ferns that are also found elsewhere in the world are *indigenous*. In the millions of years since the Hawaiian Islands came into being, isolation allowed many of the original ferns to evolve into distinct species found nowhere else in the world. These unique native ferns are called *endemic* and, as with populations of other terrestrial organisms in Hawai'i, the percentage of endemic species predominates. Of approximately 172 native species of ferns and fern allies growing in Hawai'i, 20 percent are indigenous and 65 percent are endemic. On continents, where intermixing of individuals occurs freely, the rate of endemism is much lower.

Since the arrival of humans, many other ferns have been introduced to Hawai'i. Twenty-six of these introduced ferns have escaped from cultivation and become *naturalized*. The introduced ferns included in this book are naturalized, capable of growing and reproducing in the wild without human intervention. Their spores blow to remote regions once the sole domain of the native ferns. Naturalized ferns generally do not take over native ecosystems the way many introduced flowering plants such as guava, lantana, and blackberry do. One exception is an Australian tree fern (*Cyathea cooperi*), from Queensland. Since its arrival sometime before the 1960s, spores from ferns in gardens on Kaua'i, Maui, and O'ahu have established colonies in the surrounding countryside. Australian tree ferns are better adapted to sunnier conditions than are native Hawaiian tree ferns; they grow rapidly and thrive in disturbed areas such as roadsides. On the eastern slope of Haleakalā, Maui, over 2,000 plants grow in less than 1 square mile (2.6 km^2). Native tree ferns are an important component in native plant communities; their trunks serve as nurseries for other native plants, and the fronds provide protection from wind and sun. Few plants grow on or beneath Australian tree ferns, which now threaten to displace native plants. Concern over the rapid invasion of *Cyathea cooperi* and its potential to alter native plant communities has prompted attempts to control its spread in Haleakalā National Park.

Ferns as Food and Medicine

Hawaiians used a number of ferns for food. The starchy heart of tree ferns and *ama'u* was baked and eaten during times of famine. The leaves of *hō'i'o* and *kikawaiō* were eaten raw, and *pala* was eaten cooked and made into a beverage. Being edible is not conducive to a fern's survival, and *pala* fern today is scarce in the wild, where pigs continue to eat it.

Extracts from *uluhe* and *moa* were used as laxatives. *Moa* spores were taken for diarrhea and used like talcum powder. Hawaiians brewed a cough remedy from *laukahi* (*Ophioglossum pendulum* ssp. *falcatum*) and used *pala* to reduce bronchial and intestinal mucus. A bath of *wāwae'iole* (*Lycopodiella cernuum*) brought relief from rheumatism.

Ferns and the *Hula*

In *hula* lore, *palapalai* ferns are sacred to Laka, deity of both the *hula* and the forest. Dancers adorned themselves with *lei palapalai* and decorated *hula* altars with *palapalai*.

Ferns for *Lei* Making

Lei makers weave colorful seeds, berries, and leaves using *palapalai* fronds as a backing, by adding decorative sprigs into a braid of smaller frond divisions. Hawaiians twined *palaʻā* fern with *maile* vine as offerings to their gods. Among the six basic techniques of *lei* making, ferns are most commonly used in the *hili* and *wili* methods. In *hili*, the strands of ferns are braided together without the use of thread. Ferns provide the backing for *wili lei* on which other materials are bound with thread. Pieces of *lauaʻe* are strung between segments of bright orange pandanus fruit in a third style of *lei*, called *kui*. Besides their use as personal adornment, fern *lei* also afford sun protection.

Other Uses

Hawaiians embalmed their dead with the silky *pulu* from tree ferns and *Sadleria* ferns. This cottony material was also used for bandages on wounds, and in the nineteenth century a short-lived industry exported *pulu* for stuffing mattresses.

In religious ceremonies *ʻamaʻu* was used to cover the ground where *aliʻi* walked. Hawaiians paved trails through swampy areas with *hāpuʻu* trunks. The ridgelines and corners of traditional Hawaiian thatched huts were finished with *ʻamaʻu* to ensure waterproofness, and when *pili* grass was not available this fern was substituted for thatching. Ferns found their way into native farming practices as mulch in dry regions.

The wiry black stems of maidenhair ferns were plaited into baskets and woven articles. *Kapa* makers glued strips of bark cloth, called *kapa*, together using the sap of young *ʻamaʻu* fern shoots. The outer trunk of *ʻamaʻu* fern and fronds of *palaʻā* yielded a red dye for *kapa*. When stored, *kapa* was layered with *lauaʻe* fronds to scent it.

HOW FERNS DIFFER FROM OTHER PLANTS

Some plants called ferns are not true ferns; for instance, in Hawai'i, fern trees (*Filicium decipiens*) and asparagus fern are not ferns at all. True ferns are primitive plants, which do not develop flowers or seeds for reproduction. Instead, ferns reproduce in an intriguing manner, starting with dustlike spores. Spores form in capsules called *sporangia*, which are usually found on the underside of the fronds. Some ferns have spores on all of their mature fronds, but other types have spores only on special (fertile) fronds. The obvious brownish to blackish clusters on the undersides of fronds are collections of spore capsules, called *sori*. Often the sori are protected by a flap, called an *indusium*, which varies in shape in different types of ferns. Less-evolved ferns lack this protective flap.

Spores do not simply sprout into new ferns. Instead they take a fascinating detour. The spores grow into small, heart-shaped plants called *gametophytes*, which are seldom seen because they are only a fraction of an inch across. The *gametophytes* in turn produce the eggs and sperm that combine to produce the ferns we see. Not all ferns reproduce solely through spores and *gametophytes*. Some, such as *Asplenium normale*, *Asplenium lobulatum*, and *Deparia prolifera*, produce tiny fern plantlets (*keiki*) along their fronds that take root as the aging parent frond bends to the ground.

Fern-like plants first appeared about 350 million years ago. By comparison, algae are more primitive than ferns and have existed for 500 million years. Unlike primitive plants such as mosses and algae, ferns have veins to transport materials to various parts of the plant. Figure 1 shows the relationships of ferns. This book also includes three kinds of fern allies, such as the club mosses, which have smaller leaves, simpler vein structures, and spores that often form in cones or spikes, rather than being scattered on the fronds as with ferns. Three hundred million years ago plants like ferns and club mosses dominated the world. Imagine walking through a carpet of ancestral ferns growing in a forest of club mosses 100 feet (30 m) tall. It was not until about 135 million years ago that plants evolved flowers and fruit.

FERN LIFE-STYLES

Ferns have different strategies for success (Figure 2). Ferns grow on trees, by riverbanks, in shade, and on windswept lava flows. Ferns get a head

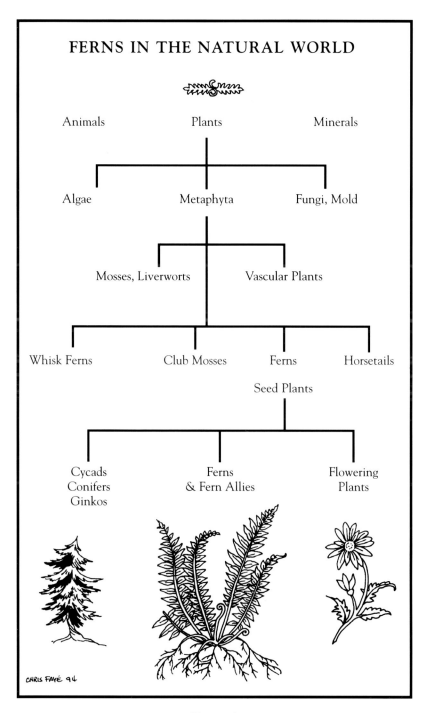

Figure 1

start above the forest floor by growing on rocks or as *epiphytes* on trees. Tree ferns stand above the competition of lower forest plants, spreading their fronds above tall trunks to gather sunlight. *Uluhe* scrambles upward to get to sunshine, and epiphytic ferns take advantage of trees to reach the light. *Terrestrial* ferns often require the shade of other trees and damp earth. Young fronds of *'ama'u* employ another survival mechanism: their reddish color seems to protect them from the sun. Ferns in similar habitats often share similar designs, textures, or growth habits.

Elevation is an important part of a fern's life-style. Some inhabit a wide range of elevations, but others occur only in a narrow range, be it lowlands, middle elevations, wet forests, or subalpine. In Hawai'i, rainfall is often a function of elevation, and many ferns have specific requirements: *Cyclosorus interruptus* grows only in boggy areas; delicate filmy ferns need high humidity; and cliff brake thrives in higher, dry areas. Because precipitation is greater on the windward (northeast) side of islands in Hawai'i, moisture-loving plants can generally be found at lower elevations on windward slopes than in leeward areas. Information on elevation ranges and moisture needs is included in the habitat descriptions. Because the majority of these ferns are abundant, they can be found readily using this information and the trails recommended at the end of this chapter.

DESIGNS IN NATURE

Ferns draw their beauty not from bright flowers and tasty fruit but from their range of patterns, from simple to complex. To appreciate this aspect of ferns, take note of how many different patterns you see in ferns along a trail. No longer will ferns look the same to you.

Patterns

The most obvious difference between ferns is how finely divided their fronds are. Some ferns, with descriptive names like stag's tongue, Māui's paddle, and adder's tongue, grow as single blades. Ferns that are once divided have simple-looking fronds branched into two rows of leaflets (*pinnae*) along a central stem. Ferns that are divided more than once repeat this same design, with leaflets that look like miniature fronds. If the smallest divisions do not indent to the midrib, they are called lobes. Figure 3 shows examples of different fern designs.

branched spikemoss (see page 22) Photo by Robert Hobdy

Christmas tree club moss,
wāwae'iole (see page 19)
Photo by David S. Boynton

puapuamoa (see page 25)
Photo by Robert Hobdy

uluhe (see page 26) Photo by David S. Boynton

uluhe growing in *'ōhi'a* at Kalalau (see page 26) Photo by David S. Boynton

uluhe lau nui (see page 28)　　　　　　　　　　　Photo by David S. Boynton

Figure 2

Textures

Another component of fern design is a frond's texture. Some ferns, like hairy stag's tongue and downy wood ferns, have minute hairs to hold moisture and possibly to discourage hungry insects. Other ferns, like *'ama'u* and bracken ferns, are thick and leathery and stand up better to the drying wind. The thin fronds of filmy ferns, some only one cell thick, need high humidity and constant moisture to survive.

SPORES INSTEAD OF FLOWERS AND FRUIT

On the undersides of their fronds, ferns have different arrangements of spore clusters: some form on the edge of the leaf, others in the middle, and some on the tips. A frond will not have spores if it is not mature or not fertile (some ferns bear spores only on certain fronds). The clusters of spore capsules (*sori*) are arranged differently in different fern families. Ferns in the same family may have varied frond designs but often share the same sorus arrangement or shape. This helps botanists to classify ferns into families and *genera*. The protective flap (*indusium*) over the sori, if one is present, can be round, oblong, or elongated; shaped like a kidney or crescent; or like an umbrella or cup. On some ferns the edge of the frond curls around to protect the sori.

WHAT DISTINGUISHES DIFFERENT FERNS

Frond shape does not necessarily dictate family relationships among ferns. When studying ferns, one can be discouraged by discovering that two ferns that look nearly identical are not even related. It may seem that scientists who grouped ferns were blind to the most obvious similarities between ferns.

I was incredulous to discover that Cretan brake (*Pteris cretica*) and sword-leaved fern (*Lindsaea ensifolia*), whose fertile fronds look so similar, belong to totally different families (Figure 4). Discouraged, I asked Tim Flynn, my fern expert, how a novice could tell these two ferns apart. "That's easy," he proclaimed, but when he picked up a magnifying lens I became suspicious about just how obvious the difference was. Would I have to crawl over the forest floor inspecting the undersides of fronds with a hand lens to understand ferns?

Ferns vary in appearance depending on their environment and the

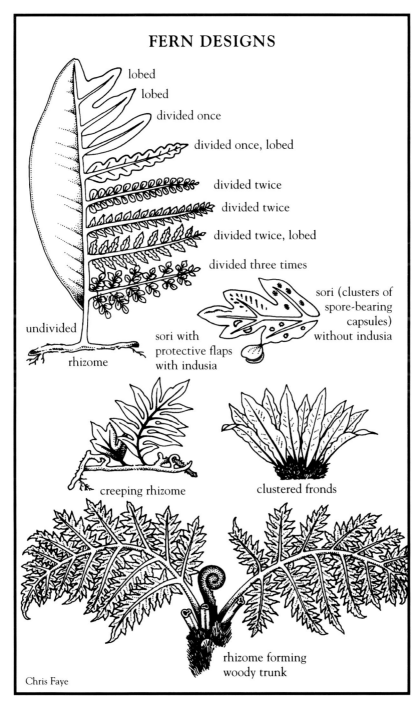

Figure 3

diversity of leaf form, so scientists classify ferns based on the arrangement of spore-bearing capsules on the undersides of fronds, a feature that changes very little. Other constant features, such as the arrangement of veins, scales, and rootlike rhizomes, also help botanists distinguish one species from another. These subtle features reveal themselves only under careful inspection aided by a hand lens.

A careful inspection of Cretan brake and lace fern's swordlike relative reveals that Cretan brake's leaf margins curl up over one side of the spores, but *Lindsaea ensifolia*'s spores lie in a marginal trough, protected by flaps on both sides. Photographs of *sori* are included in this book to help in differentiating ferns that are otherwise hard to tell apart.

This detailed differentiation is a bit daunting at first, but after one learns to recognize obvious fern features, such as frond patterns, an appreciation for finer details will evolve naturally. It is useful to carry a hand lens to appreciate the finer details of fern design; it opens a whole new dimension of enjoyment of the complexities of ferns. I find that the magnifying glass on my Swiss army knife is perfect. Soon you will find yourself peering inquisitively at the underside of ferns in urban landscaping or inspecting trailside ferns on your hands and knees to see if they are related to ferns you already know.

WHAT'S IN A NAME?

Don't worry about names at first. Even scientists do not agree entirely on what to call some ferns. Remember that names and the classification of ferns are devised by people, and no plant is more or less beautiful because of its name.

For every Hawaiian fern, there are three potential names: a common English name, a Hawaiian one, and a scientific name. The common name could be "sword fern." The scientific name is two words based on Latin or Greek and is italicized. The Hawaiian name for one sword fern is ʻōkupukupu, and *Nephrolepis exaltata* is its scientific name. With nearly 200 wild ferns in Hawaiʻi, this would mean knowing close to 600 names—an awesome undertaking. But depending on who you are talking with, whether a fellow hiker, *hula* dancer, or botanist, you will probably use different names. And if you know only one name for each fern, this book will help you cross-reference the other names.

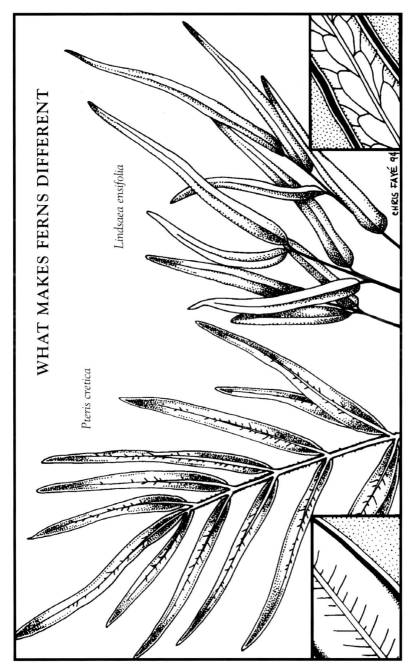

Figure 4

Scientific Names

Common names are used where possible, because they are the easiest to remember, but many ferns, especially those that were not used by the Hawaiians or that grow in obscure locations, lack both common names and Hawaiian names. If a plant could not be eaten or used as medicine, or was not particularly striking, or was too small to notice, it often went nameless until a botanist identified it. Hence scientific names are included in this book. Often, differences between closely related or similar-looking ferns are so subtle that only botanists concerned with details would give them separate names; a casual observer would lump them under the same common name. Several types of ferns, with different scientific names, might, for example, be called "maidenhair" or in Hawaiian, 'iwa'iwa.

Many people find scientific names cryptic, because they are based on Greek or Latin. Scientific names, when translated, however, often provide information about the plant they describe, so translations of these names are included in the text for added value. Because common names tend to vary around the world, and often are used for totally different plants in different locations, scientific names provide an identity for plants that remains constant across borders and language barriers. This consistency is helpful to world travelers who are keen to see what relatives of plants they know at home grow in the area they are visiting.

Like people's names, scientific names include one that describes the group (*genus*) that the organism belongs to and another that gives the organism's individual identity, but the order of the names is reversed. With scientific names the *genus* name is first, followed by the *species* name, which is sometimes followed by a subspecies (ssp.), race, or variety (var.) name. For example, the scientific name for the native hairy stag's tongue fern is *Elaphoglossum hirtum* ssp. *micans*.

The term *species* is used widely, although often people do not really know its precise definition. That is not surprising, because even taxonomists (scientists who give names to organisms) disagree on the definition of a species. One definition is that plants or animals that can breed only among themselves belong to the same species. Another definition of species is a group of individuals that, over generations, shares consistently similar features that differentiate them from others.

A *genus* is a group of species that shares less specific traits. Plants of different *genera* are further grouped into families, which share even broader similarities than do genera. For instance, the three genera of Hawaiian false staghorn ferns belong to the same family, *Gleicheniaceae*. All share the distinctive trait of repeatedly forking stems, but the ferns in one genus, *Dicranopteris*, have leaflets only along the ends of the branched stalks, the stalks of ferns in the genus *Sticherus* are leafy along their entire length, and the genus *Diplopterygium* includes species with stems that fork only once or twice.

There is some debate concerning the taxonomy and nomenclature of Hawaiian ferns. The scientific names in this book follow Charles Lamoureux's (1988) checklist of Hawaiian ferns. Where applicable, alternative names are also given, in parentheses. The taxonomy of Hawaiian ferns is currently being studied at the Bishop Museum and the University of Hawai'i by Warren H. (Herb) and Florence Wagner.

Hawaiian Names

The Hawaiian names are included in this book, because not only do people in Hawai'i often refer to ferns by their Hawaiian names, but they also give some insight into how Hawaiians saw their natural world. Hawaiian names, like the scientific name, are often descriptive. The large bladelike frond of one fern (*Elaphoglossum alatum*) was named *hoe a Māui*, because it resembles a paddle (*hoe*) of the demigod Māui. *Puapuamoa*, an arched fern (*Ophioglossum pendulum*), gets its name from its similarity to a cock's tail feathers.

It is tempting to decipher the meaning of all the Hawaiian names and determine the significance of their literal translations, but not all names have literal translations. An analogy would be a foreigner translating "Washington" as a place where 2,000 pounds of clothes were laundered. In many instances the lore has been lost that might have given some insight into the name's meaning. Only obvious translations have been included in this book and, in some cases, the literal translations are included without any suggestion of the possible connection to the fern.

WHERE TO FIND FERNS

To see a wide variety of ferns, I recommend hiking a lowland trail and then a higher-elevation trail. Most of the ferns in this book are found along the Kalalau and Alaka'i Swamp Trails on Kaua'i. On O'ahu the Mānoa Cliffs and Pauoa Flats Trails are good places to look for ferns, and many forest ferns grow along the upper portion of Waʻahila Trail. The trail to Waimoku Falls in Maui's Kīpahulu Valley is one of the few public hiking trails on that island where ferns abound; some interesting high-elevation ferns grow along Halemauʻu Trail in Haleakalā National Park. On the island of Hawaiʻi, the Crater Rim Trail in Hawaiʻi Volcanoes National Park offers a wide range of fern habitat, from tree fern forest to cinder desert, and numerous low-elevation ferns grow around ʻAkaka Falls.

Fern Allies

CLUB MOSSES
(Lycopodiaceae)

Club mosses are found worldwide except in arid areas. Different species go by the common names "ground pine," "club moss," "ground cedar," and "fir moss." In the Carboniferous Period, before the evolution of flowering plants, huge club mosses 100 feet (30 m) tall formed great forests. Club moss spores ignite easily and have been used in fireworks, flares, tracer bullets, old-fashioned flashes for photography, and for coating pills. The scientific name comes from the Greek words *lykos*, meaning "wolf," and *podium*, meaning "foot."

Christmas tree club moss,
wāwae'iole,
Lycopodiella cernua
(*Palhinhaea cernua*)

The Hawaiian name applied to all club mosses in Hawai'i, *wāwae'iole*, means "rat's foot," which describes the appearance of this fern ally. The scientific name of the genus has a similar translation, "wolf foot." *Huluhulua'iole* is another Hawaiian name for this plant. Hawaiians boiled it for three hours and then used the water to bathe rheumatic patients.

DESCRIPTION: The plant looks like a small, densely branched

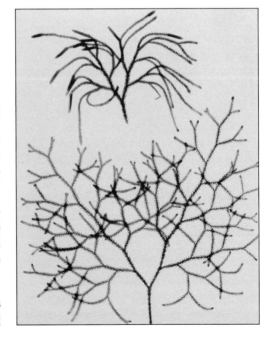

tree. The narrow branches, which fork repeatedly, are covered with tiny, pointed, pale green leaves that spiral around the stalk. Fertile cones are pale green to yellow and hang from the branch tips, hence the species name, which means "drooping." Plants can grow up to 5 feet (1.5 m) tall.

HABITAT: This plant grows on the ground in damp areas, in full sun to shade at elevations up to 4,500 feet (1,370 m).

DISTRIBUTION: Indigenous.

Huperzia phyllantha
(Phlegmariurus phyllanthus)

DESCRIPTION: This glossy, dark green club moss grows up to 2 feet (61 cm) long, often forking and ending in fertile spikes (*strobili*) 3 to 8 inches (8–20 cm) long. The species name means "leaf flowers." Leaves are longer and wider than on *wāwaeʻiole (Lycopodiella cernuum)*.

HABITAT: This plant hangs from trees and rocks in moist or wet forests up to an elevation of 2,500 feet (760 m).

DISTRIBUTION: Indigenous.

WHISK FERNS
(Psilotaceae)

Two genera make up this family of very primitive plants. The cluster of stalks in the Hawaiian genus leads to its nickname of "whisk fern." Its scientific name is based on the Greek work *psilos*, meaning "naked," because of its apparent lack of leaves.

Whisk fern, *moa*, *Psilotum nudum*

Hawaiians found this plant useful in many ways. Its spores provided a talcumlike powder for preventing chafing from loincloths and also were administered as a remedy for diarrhea in infants. A tea brewed from boiling the whole plant was used as a laxative and was also given to children with thrush.

Hawaiians devised a game with these plants. Each player made a hook from two branches of a *moa* plant and entwined them. The players pulled their branches apart much the way we break wishbones, and the winner, whose hook didn't break, crowed like a cock. This may account for the Hawaiian name *moa*, which means "chicken." Other Hawaiian names for this fern ally are *moa nahele* (*nahele* means "vegetation") and *pipi*, meaning the feelers of an insect.

DESCRIPTION: The scalelike "leaves" on whisk ferns are hardly noticeable (the species name means "naked"). Plants grow in clusters of bright green stalks as long as 2 feet (61 cm). Spores form in round, yellow clusters of two or three capsules.

HABITAT: Whisk ferns grow in rock crevices, on trees, and on the ground, up to 4,000 feet (1,220 m) above sea level.

DISTRIBUTION: Indigenous.

SPIKEMOSSES
(Selaginellaceae)

Over 500 species of small spikemosses grow worldwide, except in Antarctica. Hawai'i has two native species, and many more are grown in gardens. Elsewhere, relatives of this plant are called "resurrection plant" and "rose of Jericho." The scientific name, *Selaginella*, means a small *selago*, which was the former name for club mosses.

Branched spikemoss, lepelepeamoa, *Selaginella arbuscula*

Hawaiians call this fern ally *lepelepeamoa*, which means cock's comb. They work its rich green fronds together with roses into a *lei*.

DESCRIPTION: Plants reach over a foot (30 cm) in length and are bright green in color. Scalelike leaves form along the branches. The species name means "small, treelike." The leaves of this species are arranged in a flat plane. Spores are produced at the ends of the branches on spikelike cones up to 2.5 inches (6.4 cm) long. These plants produce two types of spore cases: one with a few large spores and the other with powderlike spores. *Lepelepeamoa* is distinctly different from the other native *Selaginella*, which grows as a single unbranched leafy strand in bogs.

HABITAT: This plant is found in damper, shadier conditions than most of the club mosses. It grows rarely as an epiphyte on other plants, but more usually on the ground or on seeping rock faces at lower to middle elevations.

DISTRIBUTION: Endemic.

Common Ferns

pe'ahi (see page 33)
Photo by Robert Hobdy

mahinalua growing with larger
'ēkaha (see page 31)
Photo by Robert Hobdy

maile-scented fern (see page 34)   Photo by David S. Boynton

golden fern (see page 34)
Photo by Kathy Valier

Venus hair (see page 41) Photo by Robert Hobdy

Gonocormus minutus (see page 47)
Photo by Robert Hobdy

hairy stag's tongue growing
with *ʻōhiʻa ku* (see page 48)
Photo by Kathy Valier

ADDER'S TONGUE
(Ophioglossaceae)

Members of this family of fleshy ferns grow in cooler climates as well as in the Tropics. Instead of spores clustering on the underside of the frond, adder's tongues are unusual, bearing their spores along two rows on a green blade that resembles a serpent (*ophis*) tongue (*glossa*), hence the scientific name.

Hanging adder's tongue, *puapuamoa*, Ophioglossum pendulum, subspecies *falcatum* (Ophioderma pendula)

Hawaiians thought this fern resembled the tail feathers (*puapua*) of a chicken (*moa*), so called it *puapuamoa*. Its scientific name describes its hanging (*pendulus*), sickle-shaped (*falcatus*) frond. Hawaiians used an infusion of *puapuamoa* as a cough remedy.

DESCRIPTION: Though not abundant, this intriguing fern is worth looking for, because it is one of the most primitive found in the wilds, looking more like some sort of hanging lily than a fern. Instead of forming its spores in clusters, it bears them on a fertile spike that dangles from the undivided frond as if glued on as an afterthought. This spike contains as many as 15 million spores. The leathery, arched fronds of this adder's tongue fern grow to almost 2 feet (61 cm) long.

HABITAT: Adder's tongue commonly grows on trees and rocks, in second-growth forest, from a few hundred feet to 2,300 feet (700 m) in elevation.

DISTRIBUTION: The subspecies is endemic. (*Ophioglossum pendulum* is indigenous.)

FALSE STAGHORN FERNS
(Gleicheniaceae)

This family of ferns carries the name of F. W. von Gleichen, a German naturalist, who lived until 1783. Most grow in or near the Tropics, but some grow as far north as Japan. Fronds are leathery with wiry stalks and grow well in full sun. The distinctive growth form of members of this family is easy to identify. Fronds are forked repeatedly into pairs of smaller divisions. Tangled masses of these ferns can reach over 20 feet (6.1 m) high, impeding hikers and smothering other plants. The spore capsules of false staghorn ferns lack a protective flap (*indusium*), which suggests that they have not evolved as far as other ferns.

False staghorn, *uluhe*, *Dicranopteris linearis*

Species in this genus grow in subtropical and warm temperate regions. Less than ten species of *Dicranopteris* exist in the world. This species is the primary invader on landslides and erosional scars. Hawaiians also called this fern *uluehe* and *unuhe*. They soaked it in water and drank the resulting liquid as a cure for constipation.

DESCRIPTION: This very common fern has fronds that fork repeatedly, ending in terminal buds. The spores form in a row of dots alongside the midrib of each division. The scientific name means "two-branched, linear." Masses of this plant can reach over 20 feet (6.1 m) high when climbing over trees and shrubs, but usually it grows only a few feet (1 m) tall.

HABITAT: This terrestrial fern takes over erosional scars, landslides, or disturbed areas from sea level to 5,500 feet (1,680 m).

DISTRIBUTION: Indigenous. It grows across the Pacific and as far as New Zealand.

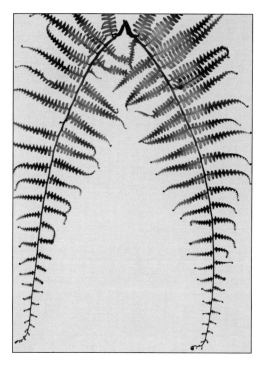

Giant *uluhe*, *uluhe lau nui*, *Diplopterygium pinnatum*

The Hawaiian name means "large-leaved" *uluhe*.

DESCRIPTION: Though this fern also grows in a tangled thicket, similar to other *uluhe*, its fronds are distinctly different: the fronds fork only once or twice along a stalk, and the divisions are much longer, paler green, and have more finely divided leaflets. The scientific name means "two-winged, divided once." Each frondlike division is about 2 feet (61 cm) long, supported on stalks 3 feet (90 cm) long. Giant *uluhe* grows in tangles several feet (1 m) thick.

HABITAT: A terrestrial fern, *uluhe lau nui* occurs on mountain slopes at middle elevations, often mixed with *Dicranopteris*, but much less common.

DISTRIBUTION: Indigenous.

False staghorn, *uluhe*, *Sticherus owyhensis*

The scientific name *owyhensis* refers to the old spelling of Hawai'i.

DESCRIPTION: This fern resembles the other *uluhe* (*Dicranopteris*) except that the narrow leafy divisions (*pinnae*) continue along the entire stalk, not just along the forked divisions. Stalks have small, stiff, black hairs. Yellow-green to brown spore clusters are arranged as in the other *uluhe*. The genus name means "lines [of sori]." This false staghorn fern is somewhat larger and sprawls more than *Dicranopteris linearis*. Fronds are a dull, dark green and reach 7 feet (2.6 m) in length with branches 3 to

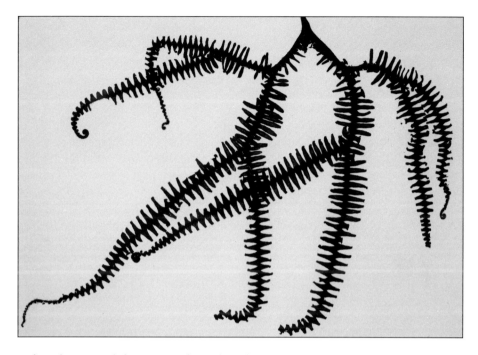

4 feet (1–1.2 m) long. It is less abundant than *Dicranopteris* and does not grow at lower elevations.

HABITAT: This is a terrestrial fern found on mountain ridges and eroded scars at elevations from 2,000 to 6,000 feet (610–1,830 m).

DISTRIBUTION: Endemic.

FINGER FERNS
(Grammitaceae)

These miniature ferns grow in damp conditions. Spore capsules form in round or oblong clusters without a protective flap *(indusium)*. All the members of this family growing in Hawai'i except one are unique to the Hawaiian Islands.

Pai, Adenophorus hymenophylloides

Pai means "to encourage or urge" in Hawaiian. Other Hawaiian names for this fern are *huna palai* and *palai huna; huna palai* means

"to hide." The scientific name means "gland bearer, like *Hymenophyllum* (a fern genus)."

DESCRIPTION: The delicate fronds are divided into two rows of leaflets, each scalloped into minute segments along the top edge and curved along the lower margin. The fronds are 2 to 6 inches (5–15 cm) long and always hang in dense clusters.

HABITAT: These ferns grow on trees and rocks in moist areas above 2,000 feet (610 m) elevation.

DISTRIBUTION: Endemic.

Palai lā'au, Adenophorus pinnatifidus (*Oligadenus pinnatifidus*)

The Hawaiian name means "fern plant." The scientific name means "gland bearer, once lobed."

DESCRIPTION: Narrow, deeply cut, light to medium green fronds reach 4 to 8 inches (10–20 cm) in length. Round *sori* lie along the middle vein of each frond division.

HABITAT: This fern grows on moss-covered trees and rocks in wet forests or along streambeds at elevations ranging from 300 to 4,500 feet (90–1,370 m).

DISTRIBUTION: Endemic.

Wahine noho mauna, Adenophorus tamariscinus var. tamariscinus
(Adenophorus tamariscinus)

The Hawaiian name for this dainty fern means "mountain-dwelling woman." The genus name means "gland bearer," and the species name refers to the tamarisk, a plant with a feathery appearance.

DESCRIPTION: The dark green fronds are divided twice and reach up to 12 inches (30 cm) in length. The lower stalk is dark brown, changing to a pale green midrib. Spores form in round dots near the tips of segments.

HABITAT: These ferns grow in clusters on trees or on the ground in shady, wet regions. In the wet areas they grow as low as 200 feet (61 m) elevation and are abundant in wet forests to elevations of over 4,000 feet (1,220 m).

DISTRIBUTION: Endemic.

Mahinalua, Grammitis tenella

Hawaiians also called this fern *kolokolo*, which means "to creep" and refers to any creeping vine.

DESCRIPTION: Like other finger ferns, *mahinalua* grows 2 to 6 inches (5–15 cm) long, but unlike its relatives, its fronds are undivided,

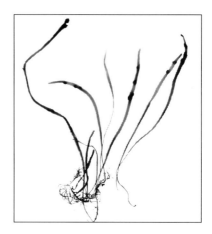

simple blades. Fronds usually appear in clusters, arising from a tangle of rhizomes. The round clusters of spores look disproportionately large compared with the slender fronds for which it gained its species name, meaning "slender." The genus name means "soft linear fern."

HABITAT: This fern is a common epiphyte in Hawaiian wet forests. It is most abundant between the elevations of 1,000 feet (305 m) and 5,000 feet (1,525 m).

DISTRIBUTION: Endemic.

Kihe, *Xiphopteris saffordii* (*Lellingeria saffordii*)

Hawaiians also called this fern *kihi*, meaning "sharp point of a leaf," possibly referring to the angular serrations of the frond. The genus name means "sword fern." William Edwin Safford was an American botanist who lived from 1859 to 1926, serving for a time as the lieutenant governor of Guam.

DESCRIPTION: The small fronds grow 2 to 5 inches (5–13 cm) long. The zigzag pattern of the divisions on the frond resembles a jigsaw blade. Spores cover the entire underside at the tips of the pinnae. This fern is not abundant, but is readily found by a careful observer.

HABITAT: *Kihe* grows in clusters on trees in wet forests from an elevation of 2,000 feet (610 m) to over 6,000 feet (1,830 m).

DISTRIBUTION: Endemic.

POLYPODY FERNS
(Polypodiaceae)

Species from this, one of the largest families of ferns, grow worldwide. Their smooth-margined fronds sprout up from a stem that either creeps along the ground or clings to trees or rocks. The name refers to the branches of their rhizomes, which are like many (*poly*) feet (*pod*). Spores form in round clusters on the undersides of the leaves.

Pe'ahi, *Microsorum spectrum*

The Hawaiians named this native fern *pe'ahi,* meaning "fan," possibly because of its shape. The scientific name means "apparition [with] small heaps of spores."

DESCRIPTION: The dark, glossy fronds are peppered with sporangia on the undersides. Fronds develop from one to five pointed lobes with a delicate network of veins and can reach up to 2 feet (61 cm) long. *Pe'ahi* is not common, but its unique design and status as the only native *Microsorum* make it worth noting.

HABITAT: This fern is terrestrial and epiphytic. It grows in wet valleys up to 2,500 feet (760 m) above sea level.

DISTRIBUTION: Endemic.

Hare's foot fern, *laua'e haole, Phlebodium aureum*

There are approximately ten species in the genus *Phlebodium*, which is named for its many veins. Though called *laua'e haole,* meaning "foreign *laua'e*" in Hawaiian, this fern reached Hawai'i a few years

before *laua'e* (*Phymatosorus scolopendria*) did. The species name means "golden."

DESCRIPTION: Fronds grow to nearly 4 feet (1.2 m) tall and are deeply lobed with conspicuous spore clusters. They are a dull green with a bluish cast, with round to oval spore clusters. Fronds sprout from creeping rhizomes, which are covered in thick golden to reddish hairs that resemble a hare's foot.

HABITAT: These ferns grow as epiphytes or terrestrially in shaded woods up to an elevation of 2,000 feet (610 m).

DISTRIBUTION: Naturalized from Tropical America.

Maile-scented fern, *laua'e*, *Phymatosorus scolopendria* P. grossus

Laua'e is mistakenly thought of as being native because of its Hawaiian name and abundance throughout all the Islands. Hawaiians adopted this introduced fern for scenting their *kapa* cloth by interleaving fronds between the folded bark cloth. Sections of *laua'e* are strung between pieces of *hala* (pandanus) for *lei*. Today, *laua'e* is abundantly cultivated around resorts and in gardens. *Laua'e* means "beloved" or "sweet," alluding to the fragrance of this fern, which resembles that of the *maile* vine from which Hawaiians made *lei*.

DESCRIPTION: The glossy green leaves of *laua'e* are easy to identify. They are dimpled on the underside by distinct spore clusters that appear as bumps on the top surface. The scientific name means "resembling a centipede, [with] swollen spore pocket." The fronds in young plants begin as undivided blades on a dark green, creeping rhizome, and then in mature plants develop into lobed fronds nearly 3 feet (1 m) long.

HABITAT: *Laua'e* grows readily on rocks, trees, or the forest floor, especially in disturbed areas and areas of secondary growth. It thrives from salt-swept boulder beaches to shaded moist lowland forests and on windswept ridges up to at least 2,000 feet (610 m). It is abundant throughout the Islands.

DISTRIBUTION: Naturalized from Old World Tropics.

Pākahakaha, Pleopeltis thunbergiana
(Lepisorus thunbergianus)

Ferns of this genus grow in the tropics. All have leathery fronds and round spore clusters at vein junctions. The genus name refers to the shield-shaped scales that cover the sori.

This, the only species of *Pleopeltis* in Hawai'i, has several names, including *'ēkaha 'ākōlea*. Part of its most common name, *pākaha*, means "having the quality of a stripe" and likely describes the long, narrow, undivided frond. Hawaiians must have noticed the similarity of its undivided fronds to those of the bird's nest fern (*Asplenium nidus*), which they also called *'ēkaha*. Another Hawaiian name,

pua'akuhinia, means "a pig satiated with rich food." The species name honors K. P. Thunberg, a Swedish botanist and explorer who lived from 1743 to 1828.

DESCRIPTION: Fronds are yellowish green to dark green with whitish undersides and curl up in dry weather. They can reach 15 inches (38 cm) in length. Spores form in oval or round cinnamon brown pockets.

HABITAT: A widespread terrestrial and epiphytic fern, pākahakaha grows in moderate shade to full sun in low wet forests and dry forests.

DISTRIBUTION: Indigenous.

'Ae, Polypodium pellucidum

DESCRIPTION: The dark green, semiglossy fronds of this fern grow to 2 feet (61 cm) in length and are divided once. The orientation of the pinnae varies to adapt to different habitats: The fronds of plants growing in forests form a single flat plane, but the variety *vulcanicum* that grows in crevices of windswept lava flows feathers its pinnae horizontally. 'Ae is among the first life forms to settle on new lava flows after lichens. Spore clusters form along the center of each pinna. Held up to the light, it has curious translucent veins that can be seen among the normal veins. The scientific name means "transparent [with] many feet." 'Ae is less common than most other polypody ferns.

HABITAT: This adaptable and variable fern grows as an epiphyte and on the ground from soggy mountaintops to desolate lava flows. It occurs from an elevation of a few hundred feet (50–100 m) in wetter regions to 6,500 feet (1,980 m) on lava flows.

DISTRIBUTION: Endemic.

CLIFF BRAKE FERNS
(Sinopteridaceae)

'Iwa'iwa, Doryopteris decipiens (Cheilanthes decipiens)

Doryopteris ferns are found around the world, with the greatest abundance in Brazil. The spear shape of some species led to its genus name, meaning "spear fern." The species name means "deceptive." Two of the thirty-five species in this genus grow in Hawai'i. The fronds of the Hawaiian species have an inverted heart shape and are smooth and leathery. Their dark, glossy stalks resemble those of maidenhair ferns, which Hawaiians called by the same name, 'iwa'iwa. *Doryopteris* ferns are also known as *manawahua* and *kumuniu*. Some botanists now classify these ferns in the family Pteridaceae.

DESCRIPTION: Fronds are medium dull green and divided twice. A dark brown line of *sporangia* forms along the frond edge. These ferns grow in clusters to 15 inches (38 cm) tall and curl up when dry. The stalk is wiry and black. Not an abundant fern, it is readily found in its favored habitats.

HABITAT: This fern thrives in exposed, often rocky sites. It grows at elevations from 500 (150 m) to 3,000 feet (915 m), either on dry ground or in crevices in rock faces, both in sun and shade.

DISTRIBUTION: Endemic.

Cliff brake, *kalamoho*, *Pellaea ternifolia*

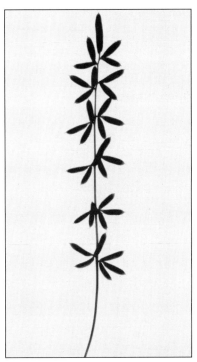

Members of this genus have dark stalks, which leads to the name *Pellaea*, meaning "dusky." South Africa and South America have the greatest concentration of the eighty species in this genus, which occurs throughout the world. Spores on all *Pellaea* ferns form linear patches along the frond margin, which curls under to protect the spores.

Literally, the Hawaiian name means "to loosen" (*kala*) and "to unfold leaves" (*moho*) and is descriptive of how the frond's divisions curl up.

DESCRIPTION: Though not abundant, the distinctive form of this fern makes it easy to recognize. Its narrow 1/2-inch (1.25 cm)-long leaflets (*pinnae*) form in clusters of threes along the glossy, dark brown stalk of cliff brake. The species name means "three-leaved." The gray-green fronds grow to 12 inches (30 cm) long.

HABITAT: This is a terrestrial fern usually found in dry open areas, but also occurring in shaded gullies, between the elevations of 1,000 and 8,000 feet (305–2,440 m).

DISTRIBUTION: A New World fern indigenous in Hawai'i.

PTERIS FERNS
(Pteridaceae)

Pteris literally means "ferns." Members of this family are usually terrestrial. Spore capsules are clustered in lines parallel to the edge of the fronds or cover most of the underside. The spores of the two species discussed here are protected by the curled leaf edge.

Cretan brake, ʻōali or ʻōwali, Pteris cretica

Because this fern was first described from Crete, its scientific name means "fern of Crete." Likely the Hawaiian name comes from the pointed (ʻō) shape of the fronds and *wali*, which means "supple and limber."

DESCRIPTION: This species produces spores on separate fertile fronds. Fronds grow to 2 feet (60 cm) tall, with wider sterile fronds. This fern is not abundant. Cretan brake commonly hybridizes with *Pteris irregularis* to form *Pteris* X *hillebrandii*.

HABITAT: This terrestrial fern can be found in dry or wet forests up to an elevation of 3,000 feet (915 m).

DISTRIBUTION: Indigenous.

Waimakanui, Pteris excelsa

The Hawaiian name *waimakanui* means "large tears" and is shared by another, rather different-looking fern, *Pseudophegopteris keraudreniana*. The scientific name means "tall fern."

DESCRIPTION: The smooth, bright green fronds grow to over 6 feet

(1.8 m) long and are divided twice. Spores form in lines under the leaf edge, which curls over to protect them.

HABITAT: This fern is terrestrial and found on shady valley floors in higher forests around an elevation of 3,000 to 4,000 feet (915–1,220 m).

DISTRIBUTION: Indigenous.

Vandenboschia cyrtotheca (see page 50)   Photo by Robert Hobdy

golden *pulu* of *hāpuʻu* (see page 51)
Photo by Kathy Valier

hāpuʻu iʻi (see page 52)
Photo by Kathy Valier

hāpuʻu frond with whitish underside
(see page 53) Photo by Kathy Valier

palapalai (see page 54)
Photo by Robert Hobdy

lace fern (see page 56)
Photo by Kathy Valier

'ākōlēa (see page 64)
Photo by Robert Hobdy

MAIDENHAIR FERNS
(Adiantaceae)

These delicate ferns grow in wet places, often clustered under a safe ledge behind a plummeting waterfall where their wiry black stalks jostle in the rush of water and air. Their water-shedding fronds lead to the name *Adiantum*, from the Greek word *adiantos*, meaning "unwetted." At least 200 species of maidenhair ferns grow around the world. Of the three types growing wild in Hawai'i, two are naturalized, and one is native to the Islands. Some botanists now classify maidenhair ferns in the family *Pteridaceae*.

The Hawaiians called all maidenhair ferns 'iwa'iwa. The root word, 'iwa, comes from the Hawaiian word for frigate bird. A chant and *hula* likens the dancers to the gliding, soaring, and dipping grace of the 'iwa birds, much like a bank of 'iwa'iwa ferns waving in the breeze.

Venus hair, 'iwa'iwa, Adiantum capillus-veneris

Hawaiians plaited the distinctive shiny black stalks, adding them to tan mats made of *lauhala* (pandanus) to create decorative designs. The species name means "fine as Venus hair."

DESCRIPTION: At first glance this fern looks like the common (introduced) maidenhair, but it has deeper divisions between each lobe, and the spore pockets are rectangular, not round. Leaves measure 3/4 inch (19 mm) across—a little larger than each of the common maidenhair's numerous subdivisions. The fronds of both

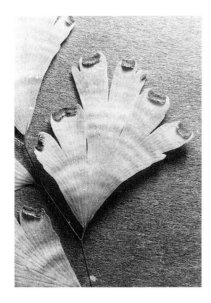

species grow to 16 inches (40 cm) long.

HABITAT: This native fern grows in damp to wet conditions on shaded, moist rock faces. It is also common in areas too harsh for the common, introduced species and grows along coasts, where it is subject to salt spray. It is found at lower elevations, below 2,000 feet (600 m) and is much less common than the introduced maidenhair ferns.

DISTRIBUTION: Indigenous.

Common maidenhair, 'iwa'iwa, *Adiantum raddianum* (*Adiantum cuneatum*)

This fern, the most common wild maidenhair in Hawai'i, was introduced from the Eastern Hemisphere. Hawaiians used it in the same ways that they used native maidenhair fern and called both by the same name. The scientific name for this species commemorates Giuseppe Raddi (1770–1829), an Italian botanist.

DESCRIPTION: Fronds of this maidenhair reach 10 to 16 inches (25–40 cm) in length and are pendent. The round spore pockets and more delicately subdivided leaves distinguish it from the native maidenhair, *Adiantum capillus-veneris*.

HABITAT: This fern grows in very wet conditions; it is found on rock cliffs dripping with water, in disturbed areas in native forests at middle to lower elevations.

DISTRIBUTION: Naturalized.

Five-finger maidenhair, *Adiantum hispidulum*

DESCRIPTION: This maidenhair produces curved sprays of pale green to deep green fronds that reach 2 feet (61 cm) in length, with each of the fingers in its handlike frond reaching several inches (20 cm) long. New fronds are often maroon in color. Spores form in kidney-shaped spore pockets. This species is also called rough maidenhair. The species name means "bristly."

HABITAT: This fern grows along the banks of intermittent streams and on hillsides, in drier conditions than the common maidenhair, with only occasional moisture and shade. Five-finger maidenhair can be found growing as high as 3,000 feet (915 m) above sea level.

DISTRIBUTION: Introduced around 1910 and now naturalized.

SILVER FERNS
(Hemionitidaceae)

Bamboo fern,
lo'ulu, *Coniogramme pilosa*

Most of the twenty species in this genus come from Malaysia, and the rest range from Japan to Africa. The Hawaiian name, *lo'ulu*, could mean "like a breadfruit," referring to the tree, which has somewhat similarly shaped leaves. The arrangement of its spore capsules inspired the scientific name of this genus, *Coniogramme*, meaning "dust line." The species name means "[with] shaggy soft hairs." Some botanists classify silver ferns in the family *Pteridaceae*.

DESCRIPTION: *Lo'ulu*'s smooth fronds can reach 3 feet (1 m) in length and are divided once. Leaflets (*pinnae*) have finely serrated edges and reach 6 inches (15 cm) long and 1 inch (2.5 cm) wide, tapering at each end. *Sori* form in prominent lines along veins, angling out from the *pinnae* midribs. The stalk is smooth and pale yellow.

HABITAT: This fern grows on the ground in dry to wet, shady forests from an elevation of 440 feet (134 m) to over 5,000 feet (1,525 m).

DISTRIBUTION: Endemic.

Golden fern, silver fern, *Pityrogramma calomelanos*

About fifteen species of the genus *Pityrogramma* have been described in the world, mostly from the Americas, but they also grow in Africa, Madagascar, and the Mascarene Islands.

Some sources give this fern the Hawaiian name of *laukahi*. On Niʻihau it is referred to as *palapalai*.

DESCRIPTION: This abundant fern has a golden or white powder

covering the underside of its fronds, making it easy to identify. Fronds are divided twice, and the smallest divisions *(pinnae)* are slightly scalloped. The genus name means "husk line" and describes the arrangement of the buff-colored spore capsules, which form along the division veins. The species name means "beautiful, dark." The dark green fronds are smooth and stiff, and their dark brown stalks rise from tangled rhizomes. The entire frond can reach 3 feet (1 m) long.

Some botanists consider golden fern, with its curry-colored undersides, a separate species *(Pityrogramma austroamericana)*. Slapping either of these ferns on a piece of dark paper or a teeshirt makes an attractive print.

HABITAT: A hardy terrestrial fern, golden and silver ferns grow in the open or in shade in all but the driest regions, from sea level to 4,000 feet (1,220 m) elevation.

DISTRIBUTION: Introduced around the year 1903 and now naturalized.

FILMY FERNS
(Hymenophyllaceae)

The translucent fronds of filmy ferns make them distinctive in the fern world and led to this family's scientific name, which describes its membrane ("hymen")-like fronds. Because these tender ferns dry out easily, they grow only in deep shade in wet forests and curl up during dry spells. The spore clusters on filmy ferns form along the frond edges sandwiched between two flaps or in tubes. It is easy to recognize filmy ferns because of their translucent fronds, most of which are only one cell thick.

Gonocormus minutus

This is the smallest of native ferns. The scientific name means "minute [fern that] generates [from] bulb."

DESCRIPTION: This unusual tiny, fan-shaped fern easily can be passed over because it looks like a mat of moss growing on rocks or

trees at low elevations. Often its 3/8-inch (9 mm)-wide jade-colored fronds grow intermixed with mosses. By growing close together, these ferns shield each other and preserve moisture. Inspecting this diminutive fern with a magnifying lens reveals the trumpet-shaped cups that form along the fern's edge, bearing its spores.

HABITAT: This fern grows on moss-covered boulders and the bottoms of tree trunks in moist, shaded areas at elevations of up to 2,000 feet (610 m).

DISTRIBUTION: Indigenous.

'Ōhi'a kū, *Mecodium recurvum*

The Hawaiian name, *'ōhi'a kū*, means "standing [on] *'ōhi'a*" and aptly describes this epiphytic fern of Hawaiian wet *'ōhi'a* (*Metrosideros*) forests. The scientific name means "curved length backwards."

DESCRIPTION: Fronds are pendent, highly divided, and lacy looking and reach up to 8 inches (20 cm) in length. Their telltale translucence identifies them as filmy ferns. Spores form in clamshell-shaped *indusia* imbedded in the pinnae tips. Fronds rise from threadlike creeping rhizomes. This is one of the most abundant filmy ferns found in the forest.

HABITAT: This fern grows on moss-covered trees in wet forests from an elevation of 2,100 feet (640 m) to an elevation of over 4,500 feet (1,370 m).

DISTRIBUTION: Endemic.

Palai hinahina,
Sphaerocionium lanceolatum

Hawaiians named this diminutive fern "white-haired *palai*" after the lacy *palai* fern. The scientific name means "globe-shaped, lance."

DESCRIPTION: The fronds average around 4 inches (10 cm) in length and are hairy, but not as hairy as its close relative, *Sphaerocionium obtusum*. Spores form in round, clamshell-shaped *indusia* encased at the pinnae ends.

HABITAT: This fern grows as a mat-forming epiphyte in wet forests from 2,000 feet (610 m) to over 4,000 feet (1,220 m) in elevation.

DISTRIBUTION: Endemic.

Vandenboschia cyrtotheca

Three species of this genus grow in Hawai'i. *Vandenboschia cyrtotheca* closely resembles *V. davallioides*, which the Hawaiians called *palai hihi*. The Hawaiian name might have applied also to *V. cyrtotheca*, although this dual usage is not recorded. The scientific name of this species means "Vandenbosch's [fern with] curved containers."

DESCRIPTION: The large, dull jade-green fronds reach about 16 inches (40 cm) in length and are translucent like fronds of all other filmy ferns. The fronds are triangular and divided twice, with rough edges on the smallest divisions. Trumpet-shaped *indusia* con-

tain the spores. This species differs in shape from *V. davallioides*, which has elongated diamond-shaped fronds. It is not an abundant fern.

HABITAT: These ferns grow on shaded, dripping cliff faces and stream banks or on mossy rocks, in either native or second-growth forests from sea level in wet locations to 4,000 feet (1,220 m) in elevation.

DISTRIBUTION: Endemic.

TREE FERNS
(Dicksoniaceae)

Aerial roots of these ferns mass together in a treelike trunk. Fronds arise, free of scales, from a mass of woolly hairs, which differentiates these tree ferns from "Australian" tree ferns (Cyatheaceae) which bear scales, although some botanists now include Hawaiian tree ferns in the Cyatheaceae family. Longer, finely divided fronds and taller stature distinguish them from *Sadleria* ferns, which also form trunks. Spores cluster in little satchels at the ends of small veins, often along the edges of fronds. In New Zealand some tree ferns in this family grow up to 50 feet (15 m) tall. The family name honors a British botanist, James Dickson, who lived until the late 1800s.

In Hawai'i this family is represented by only one genus, *Cibotium*. Ferns in this genus also grow in Southeast Asia, in Malaysia, and in Central America. The name *Cibotium* comes from the Greek word for box or satchel, *kibotos*, referring to the boxlike *indusia*. Four species of *Cibotium* grow in Hawai'i.

The *pulu* was used to embalm dead bodies, which could be preserved for up to eight months before burial. Later, in the latter part of the nineteenth century, *pulu* was shipped to California for stuffing mattresses and pillows. Between 1867 and 1884 over 2,200 tons of *pulu* from tree ferns and *Sadleria* ferns was exported from Hawai'i.

The starchy heart of tree ferns provided food for Hawaiians during times of famine for fattening their pigs. In the 1920s a short-lived operation on the island of Hawai'i produced starch for laundry and food use from *hāpu'u*. Hawaiians also ate the young fiddlehead fronds, cooked. The trunks were once used by Hawaiians to pave trails through boggy areas.

Today, *hāpu'u* is popular in gardens as an ornamental plant and the trunks are used as a medium for growing orchids. In the wilds, seedlings of other plants often begin as epiphytes growing on the trunks.

Hawaiian tree fern, *hāpu'u 'i'i*, *Cibotium menziesii*

The Hawaiian name means "small *hāpu'u*." It was also referred to as simply *'i'i*. This species is named after Archibald Menzies (1754–1842), a physician and botanist aboard Captain Vancouver's ship.

DESCRIPTION: These are Hawai'i's largest tree ferns, reaching 35 feet (10.7 m) in height. Their younger fronds and bases may be covered with brown hairs, and the upper frond stalk is covered with stiff, black hairs. Fronds are smooth, with slightly paler undersides and yellowish midribs.

HABITAT: This tree fern is found in moist forests and rain forests from 1,000 feet (305 m) to 6,000 feet (1,830 m) in elevation. The plants grow both terrestrially and epiphytically.

DISTRIBUTION: Endemic.

Tree fern,
hāpu'u, Cibotium glaucum

This fern is also called *hāpu'u pulu*. Another name, *hāpu-'upu'u*, refers to those ferns too young to have developed trunks.

DESCRIPTION: The smooth fronds are distinctly whitish underneath (the species name means "grayish"). The soft golden *pulu* extends up the fern stalk, instead of gradually becoming black, stiff hair as on *Cibotium menziesii*. These tree ferns stand about 10 feet (3 m) shorter than *Cibotium menziesii*, reaching 25 feet (7.6 m) in height.

HABITAT: This tree fern is found in dry to damp forests up to an elevation of 6,000 feet (1,830 m), in association with *'ōhi'a* trees. It is especially abundant on the island of Hawai'i.

DISTRIBUTION: Endemic.

HAY-SCENTED FERNS
(Dennstaedtiaceae)

Species in this family grow in tropical and subtropical parts of the world. Other members of this family include "hay-scented ferns," "cup ferns," and "lady ground fern." The family name honors a

German botanist, August Dennstaedt, who lived in the early nineteenth century.

Palapalai, Microlepia strigosa

There are about forty-five species of *Microlepia* in the world, all medium to large in size. Spore capsules cluster along the leaf margin, protected by small, half-cup-shaped *indusia*.

This fern's other Hawaiian names are *palai* and *palaiʻula* (meaning "with dark stalk"). In Hawaiian, *palai* means "to turn one's face away in bashfulness or humility." *Palapalai* is important in *hula*: it is used to adorn *hula* altars. *Hula* dancers value its woody fragrance and suppleness for weaving into adornments.

DESCRIPTION: Bright green fronds grow to almost 5 feet (1.5 m) long. This common fern resembles lace fern, but its *pinnae* are not as deeply lobed as those of lace fern, and it is covered with small hairs, for which it is named (the scientific name means "[having] small scales [and] stiff bristles").

HABITAT: This is a terrestrial fern growing in moist, shady places from 750 feet (230 m) to 6,000 feet (1,830 m) in elevation.

DISTRIBUTION: Indigenous.

BRACKEN FERNS
(Hypolepidaceae)

Some botanists now include bracken ferns in the family *Dennstaedtiacea*.

Bracken fern,
kīlau, *Pteridium aquilinum* subspecies *decompositum*
(*Pteridium decompositum*)

Some botanists consider *Pteridium aquilinum* the only species in this genus, though there are many varieties. Bracken ferns are found worldwide. American Indians ate them, Russians and Germans brewed beer from the shoots, and Japanese eat both young shoots and the starchy rhizome.

This fern shares the same Hawaiian name as a stalk of shredded *ki* leaves *(lau)* used in blessings and guiding fishing canoes. Hawaiians also referred to bracken fern as *kīlau pueo* and *pai'ā*.

DESCRIPTION: The stalk seen in cross section is said to be shaped like an eagle, leading to the species name, *aquilinum*, meaning "eagle." Some attribute the name to the winglike shape of the frond divisions. The slightly glossy, bright green fronds stand 3 feet (1 m) tall and are divided three times. Stalks are stiff, shiny, and golden colored. Each division ends in a long, undivided segment. Spore clusters are not numerous and form, when present, under a protective *indusium* at the frond edge. Fronds grow along creeping subterranean rhizomes rather than in clusters.

HABITAT: *Kīlau* is terrestrial, growing vigorously in open areas or light shade. Its habitat includes barren lava fields and dry, rocky roadsides, from 550 feet (170 m) to 10,000 feet (3,050 m) elevation.

DISTRIBUTION: This subspecies is endemic.

LACE FERNS
(Lindsaeaceae)

Lace fern, *palaʻā*, *Sphenomeris chinensis* (*Odontosoria chinensis*)

This genus was named for the wedge shape of the smallest divisions (*pinnae*) of its fronds. The species name means "Chinese." The eighteen species in this genus grow in the Tropics and subtropics, with some as far from the equator as Japan and New Zealand. All *Sphenomeris* ferns have hairless fronds. Spores form where veins end at the tips of the smallest segments and escape through a slit at the outer margin of the *indusium*.

Hawaiians made a brown dye from *palaʻā* fronds, and native women sought remedies from it for various female maladies. The fine fronds are still used to decorate *hula* altars. Hawaiian mythology connects this fern with Hiʻiaka, the younger sister of Pele, the volcano goddess. Hiʻiaka shielded herself from the huge lizardlike

moʻo of Puna by wearing a skirt of *palaʻā*, then trapped those fearsome mythological creatures in a tangle of *palaʻā* ferns. Another Hawaiian name for the fern, *pāʻūopalaʻe*, refers to a woman's *pāʻū* skirt, possibly made of fronds.

DESCRIPTION: The glossy fronds grow up to 5 feet (1.5 m) long and are divided three times. Flat pockets protect the spore clusters at the ends of the segment lobes. Fronds sit atop burnt-orange stalks, which grow in clusters. This is a very common fern throughout the Islands.

HABITAT: This delicate terrestrial fern lives easily in a wide

bird's nest fern (see page 62)　　　　　　　　　　　　　　Photo by Robert Hobdy

stag's tongue (see page 71)
Photo by Kathy Valier

rasp fern with salmon-colored fronds
(see page 77)
Photo by Kathy Valier

'ama'u frond (see page 78) Photo by David S. Boynton

'ama'u (see page 79) Photo by Kathy Valier

'apu'u (see page 80)
Photo by Robert Hobdy

water fern (see page 80)
Photo by Robert Hobdy

range of situations from exposed road cuts to lush forest floors. It grows from sea level to over 4,000 feet (1,220 m) in elevation.

DISTRIBUTION: Indigenous.

MAIDEN FERNS
(Thelypteridaceae)

The scientific name of this family comes from *thelys*, meaning "female," and *pteris*, meaning "fern." The veins on fronds of these ferns extend to the leaf margin. Leaf stalks are flanked by two strap-shaped conducting bundles.

Kikawaiō, Christella cyatheoides
(*Thelypteris cyatheoides*)

The roots and fronds provided food and medicines for the Hawaiians. Other Hawaiian names include *pakikawaiō* and *kupukupu makaliʻi*.

DESCRIPTION: The dull green fronds, with whitish midribs, reach 5 feet (1.5 m) in length and are divided into pinnae with serrated edges. Round spore clusters are generally aligned in parallel rows, flanking the veins that branch off the leaflet midribs. This fern resembles *Pneumatopteris sandwicensis*, whose leaves are scalloped. On Kauaʻi the rhizome of this fern forms a trunk, which may account for its species name, which alludes to the *Cyathea* family of tree ferns. The lower stalk is pale and winged in this species.

HABITAT: This terrestrial fern abounds in areas of moderate rainfall from elevations of 600 to 4,000 feet (300–1,220 m), usually on valley floors and in side gullies.

DISTRIBUTION: Endemic.

Downy wood fern, *paiʻiʻiha* Christella dentata (*Thelypteris dentata*)

This fern grows throughout tropical and temperate regions of the world and was introduced to Hawaiʻi around 1887.

DESCRIPTION: This is one of the most common ferns in Hawaiʻi. Its soft, green, fuzzy fronds reach up to 40 inches (1.2 m) tall and are divided once into toothed *pinnae*, hence its species name, *dentata*, meaning "toothed." Spore clusters form between the middle vein and the edge of the pinna. Downy wood fern, with two to five reduced lower pinnae, often grows with its close relative, *Christella parasitica*, in which the lower pinnae taper outward.

HABITAT: This terrestrial fern grows in wet, shaded forests from sea level to an elevation of 2,000 feet (610 m).

DISTRIBUTION: Naturalized.

Swamp cyclosorus, *neke*, Cyclosorus interruptus (*Thelypteris interruptus*)

Neke is also a rare variation of the word *nakeke*, which means "rustling," and could refer to the sound of the stiff fronds of this fern rubbing together. The scientific name means "interrupted, circular spore clusters."

DESCRIPTION: The leathery, hairless fronds of *neke* distinguish it from downy wood fern. Its fronds stand nearly 4 feet (1.2 m) tall and are branched into slender, tapered divisions with coarse, sawlike edges. Their color varies from glossy yellow-green to dark green. Very few of the fronds bear spore clusters.

HABITAT: A terrestrial fern, *neke* is common in swampy areas as high as 4,000 feet (1,220 m) above sea level.

DISTRIBUTION: Indigenous.

Hō'i'o kula, *Pneumatopteris sandwicensis* (Stegnogramma stegnogrammoides)

The name of the genus means "air fern." The species name comes from the old name for Hawai'i, the Sandwich Islands.

DESCRIPTION: Fronds are divided once, with spores clustered in pockets along either side of the midrib. Dark brown scales cover the bottom of the stalk. Fronds are downy and grow to 4 feet (1.2 m) long. The fronds are similar to those of *Christella cyatheoides* except that the lobes are rounded, not pointed, and the tips taper instead of being

divided to the end. It forms a small trunk as tall as 8 inches (20 cm).

HABITAT: This terrestrial fern grows in a wide range of landscapes, from stream banks to sparsely wooded slopes, at elevations from 1,500 feet (455 m) to 5,000 feet (1,525 m).

DISTRIBUTION: Endemic.

SPLEENWORT FERNS
(Aspleniaceae)

The scientific name originates from the use of spleenworts in curing spleen diseases. There are 650 species of *Asplenium*, growing mostly in the Tropics and subtropics. Spores form under indusial flaps running along the veins of the fronds. Many of the twenty or so species in Hawai'i are variable in form, making some of them hard to differentiate.

Spleenwort, 'iwa'iwa, *Asplenium adiantum-nigrum* (Asplenium X adiantum-nigrum)

This fern shares the same Hawaiian name as maidenhair ferns, to which the scientific name, *adiantum*, also refers, probably because of the glossy dark stalk of both plants.

DESCRIPTION: The stalk changes from shiny dark brown below to green part way up the frond. Fronds grow to 16 inches (40 cm) tall and are divided twice. Spores form under elongated *indusia*, which are arranged in a chevron pattern on the undersides of the delicate fronds.

HABITAT: *'Iwa'iwa* grows terrestrially on cinder plains, lava flows, and in dry forest, ranging in elevation from 2,000 feet (610 m) to 8,000 feet (2,440 m).

DISTRIBUTION: Indigenous.

Spleenwort, *'anali'i*, *Asplenium lobulatum*

The Hawaiian name, *'anali'i*, literally means "stunted." Another Hawaiian name for the fern, *pi'ipi'ilaumanamana*, describes its "wavy-branched leaf" (*pi'ipi'ilau*) and its habit of producing young plantlets along the midrib (*manamana* means "appendages" and "to branch out"). The species name means "with small lobes."

DESCRIPTION: Leathery fronds are yellowish green and glossy above and dull gray-green below, with coarse veins. *Sori* angle out from the midrib in a chevron pattern. Fronds are divided once and grow to 32 inches (81 cm) long. *Pinnae* are either pointed or blunt. Plantlets, called *keiki* (babies) in Hawai'i, often form along the midrib of the frond near the tip.

HABITAT: This fern grows on the ground and occasionally perches on trees or rocks in deep shade, from 2,000 feet (610 m) to over 4,000 feet (1,220 m).

DISTRIBUTION: Indigenous.

Bird's nest fern, ʻēkaha, ʻākaha, *Asplenium nidus*

Bird's nest fern grows elsewhere in Polynesia and as far away as Africa. It resembles a large stag's tongue fern (*Elaphoglossum*), with which it shares the Hawaiian name ʻēkaha. Hawaiians used the black midrib of bird's nest fern in decorating woven mats in much the same way maidenhair fern stalks were used. Hawaiians also used it to cover the stump where a log had been felled for a canoe. Young or small bird's nest ferns are called ʻēkahakaha. The species name means "nest."

DESCRIPTION: The linear arrangement of spore cases along the veins is very different from the solid mass of spore cases on fertile stag's tongue fern fronds. Fronds are 2 feet (61 cm) to 4 feet (1.2 m) tall. The distinctive rosette shape of this fern is the easiest way to identify it.

HABITAT: The nest-shaped clusters of undivided fronds of this fern grow perched in trees or on the ground. Bird's nest fern grows up to an elevation of about 2,500 feet (760 m).

DISTRIBUTION: Indigenous.

Spleenwort, *Asplenium normale*

DESCRIPTION: Fronds of this fern are divided once, with elongated *sori* running parallel to the veins on each division. Fronds are a dull, dark green on top and pale below and can reach 20 inches (51 cm) in length. Tiny ferns are produced at the frond tips. Stalks are brownish black. The species name means "regularly patterned."

COMMON FERNS

HABITAT: This fern grows on the ground and on fallen trees, mostly from elevations of 1,500 to over 4,000 feet (455–1,220 m), in wet areas and deep shade.

DISTRIBUTION: Indigenous. Not found on Lāna'i.

LADY FERNS
(Athyriacae)

The name of this family means "without shield" and alludes to the lack of shields on their clusters of spore capsules. The spore capsules cluster in oblong or narrow strips flanking the leaf veins. Elsewhere, species in this family of delicate ferns include "glade ferns." Some botanists now include lady ferns in the family *Dryopteridaceae*.

Athyriopsis japonicum
(Deparia petersenii)

DESCRIPTION: This is a common, weedy fern with fronds to 32 inches (81 cm) long. Brown, furry scales cover the stalk. Fronds are dull, medium green and are covered with minute white hairs. *Sori* are arranged in a chevron pattern on the undersides.

HABITAT: This is a common terrestrial fern, usually found in the shade. It is adapted to a wide range of elevations and habitats of mainly disturbed areas and pathways in secondary growth forest to wet forest up to an elevation of 4,000 feet (1,220 m).

DISTRIBUTION: Naturalized.

'Ākōlea, *Athyrium microphyllum*

This fern shares the same Hawaiian name as the mottled periwinkle, a land mollusk. A song called *"Kāhuli Aku Kāhuli Mai Kāhuli Lei 'Ula Lei 'ākōlea"* plays on the word *kōlea*, in describing a *kōlea* bird coming to drink at a freshwater pool named *ā kōlea* on the island of Hawai'i and seeing there the reflection of the *'ākōlea* fern and snail.

DESCRIPTION: The most delicately divided of Hawaiian ferns, *'ākōlea'*s fronds are commonly divided three times, and the smallest divisions are finely lobed, leading to the name *micro* (small) *phyllum* (leaved). Bright green *'ākōlea* fronds grow on small trunks 6 to

8 inches (15–20 cm) tall. Minute hairs form on the upper side of the fronds, which reach to 42 inches (1.1 m) long. Elongated clusters of spore capsules are arranged in a chevron pattern.

HABITAT: This fern grows in wetter mountains from 2,000 (610 m) to 6,000 feet (1,830 m).

DISTRIBUTION: Endemic.

Deparia prolifera var. *marginale*

DESCRIPTION: Fronds are smooth and divided twice, with deeply lobed *pinnae*. Smooth brown stalks and fronds can grow to over 5 feet (1.5 m) long. Spores lie beneath *indusia* that are arranged in a chevron pattern and stick out from the margin of the divisions. This fern also reproduces through sprouts along the midrib of the frond and along the midrib of the pinnae. This is not an abundant fern, but worth noting for its unusual asexual mode of reproduction.

HABITAT: This is a terrestrial fern found in middle elevations from 1,000 feet (305 m) to 3,500 feet (1,065 m), usually in deep shade along streambeds.

DISTRIBUTION: Endemic. Not found on Lāna'i.

Hōʻiʻo, *Diplazium sandwichianum* (*Athyrium sandwichianum*)

Hawaiians ate the young fronds of hōʻiʻo with freshwater shrimp (ʻōpae) and, after Western contact, with salted salmon. One translation of the Hawaiian name, "acting like" (ho), "meat" (ʻiʻo), may refer to the Hawaiian practice of substituting certain greens for meat. On Maui this fern is called *pohole*. The genus name refers to the arrangement of *indusia* on both sides of the vein.

DESCRIPTION: This fern is abundant, with soft, medium green fronds sometimes reaching over 5 feet (1.5 m) in length. *Sori* are

oblong and arranged in a chevron pattern. *Hōʻiʻo* can be differentiated from *Dryopteris glabra*, a smaller fern of rather similar size, shape, and habit, whose spore capsules group in small round clusters.

HABITAT: *Hōʻiʻo* grows on shady patches of open ground in the mountain forests, usually from elevations of 2,000 (610 m) to 6,000 feet (1,830 m).

DISTRIBUTION: Endemic.

WOOD FERNS
(Aspidiaceae)

This family of ferns includes "shield ferns" and "male ferns." Some botanists now classify wood ferns as members of the family *Dryopteridaceae*. *Dryopteris* species occur worldwide. In some temperate areas they grow in oak woodlands, hence the Greek scientific name, meaning "oak fern." All *Dryopteris* ferns are good-sized ferns with crown-shaped clusters of leathery fronds. Spore capsules form round clusters on the underside of the *pinnae*. A kidney-shaped covering *(indusium)* protects the spores. One species in this genus was used as a tapeworm remedy.

Hawaiian bow fern, kā'ape'ape, *Cyrtomium caryotideum*

Another Hawaiian name for this bluish green fern, 'āhina kuahiwi, means "mountain gray." The scientific name for this species may suggest the similarity of its fronds to those of the caryota palm. Others in this genus are called "house holly fern."

DESCRIPTION: This fern is not common, but the plant is interesting because it is the only native *Cyrtomium* and is easily recognized by its distinctive shape. Its dull fronds are divided once with pointed *pinnae* curving upward. The margins are finely toothed. Fronds grow in clusters and reach 16 inches (40 cm) in length, with clusters of spores scattered on the undersides. A more common introduced bow fern, *Cyrtomium falcatum*, has thicker, more glossy, leathery fronds and grows in disturbed areas and along exposed coastal cliffs of O'ahu, Moloka'i, Maui, and Hawai'i.

HABITAT: This terrestrial fern grows in disturbed moist forests or open disturbed areas, often at an elevation of 2,000 feet (610 m) to 3,000 feet (915 m).

DISTRIBUTION: Indigenous.

'Akole, Dryopteris unidentata var. unidentata (Dryopteris unidentata)

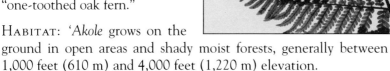

DESCRIPTION: Dull, yellowish green, triangular fronds are highly divided, with *sori* at the base of each notch between the lobes. Black hairlike scales at the base of each frond stalk (*stipe*) differentiate this fern from another similar-looking fern, Dryopteris glabra (see hō'i'o), which has flatter brown scales. Fronds grow up to 4 feet (1.2 m) long. The scientific name means "one-toothed oak fern."

HABITAT: 'Akole grows on the ground in open areas and shady moist forests, generally between 1,000 feet (610 m) and 4,000 feet (1,220 m) elevation.

DISTRIBUTION: Endemic.

Dryopteris wallichiana

DESCRIPTION: The stalk is covered with golden scales. The glossy, rich green fronds reach almost 4 feet (1.2 m) in length and are divided twice, with deeply lobed pinnae. Spores are arranged in clusters between the leaf margin and midrib at the end of small veins. This fern grows in a distinctive crown shape rising from a small trunk only a few inches (about 10–20 cm) tall. Dryopteris fusco-atra closely resembles D. wallichiana, except that its fronds are

more supple and narrow, and black, not golden, hairs cover its stalk. The scientific name means "Wallich's oak fern."

HABITAT: This fern grows in moist, shady forests between 3,000 feet (915 m) and 6,000 feet (1,830 m).

DISTRIBUTION: Indigenous.

Tectaria gaudichaudii (*Tectaria cicutaria* ssp. *gaudichaudii*)

The genus name is derived from the word *tectum*, which means "roof" and refers to the shape of the spore case covers. Charles Gaudichaud-Beaupré, for whom this species is named, was a botanist who visited Hawai'i between 1819 and 1841 aboard French ships. Ferns in this genus grow in the warm regions of both Eastern and Western Hemispheres. Most *Tectaria* form clusters of triangular-shaped blades. Spores are borne in round capsules mostly at the ends or on the sides of the veins.

One Hawaiian name mentioned by some sources is 'iwa'iwa lau nui, which means "large-leaved maidenhair," because of its dark, wiry stalk.

DESCRIPTION: The dull, dark green fronds are thin, textured, and divided once, with deep lobes under which round clusters of spore capsules form. The stalks are a glossy dark brown. Size varies from only 6 inches (15 cm) to 4 feet (1.2 m) long. This is not an abundant fern.

HABITAT: These ferns are terrestrial and grow in wet shady areas, such as gulches, up to 6,000 feet (1,830 m) in elevation.

DISTRIBUTION: Endemic.

STAG'S TONGUE FERNS
(Elaphoglossaceae)

The undivided tonguelike fronds of these ferns inspired the scientific family name, which means "stag's tongue." Members of this group thrive in warm climates, perching on trees or sprouting from the ground. Their leathery fronds are either sterile (bearing no spores) or fertile and covered with spores. Some botanists include these ferns in the family *Dryopteridaceae*.

Of the 600 species of stag's tongue ferns known, Hawai'i has nine different *Elaphoglossum* species, all except one of them unique to these Islands. Hawaiians called the larger fronds "paddle of Māui" for their demigod. Other *Elaphoglossum* share the name 'ēkaha with bird's nest ferns, which also have undivided fronds.

Botanists differentiate between the species by the pattern of their veins, frond shape, hairs, and scales. Veins either form net patterns or are parallel and may or may not connect along the frond edge.

Stag's tongue fern, *laukahi nunui*,
Elaphoglossum aemulum

The name *laukahi nunui* means "large single leaf." This fern is also called 'opeha. The scientific name means "imitating stag's tongue."

DESCRIPTION: Fronds are smooth, undivided blades, 9 to 16 inches (23–40 cm) long and 1 1/3 to 1 3/4 inches (34–44 mm) wide. Fertile fronds are smaller: 6 to 13 inches (15–33 cm) long and only 3/4 to 1 1/3 inches (19–34 mm) wide. Large scales cover the rhizome.

HABITAT: This stag's tongue fern grows on trees in moist, shady forests above 1,000 feet (305 m) in elevation, where it forms skirt

Elaphoglossum crassifolium left and middle; *E. aemulum* on right.

like clusters around tree trunks and less commonly grows on the ground and on rocks.

DISTRIBUTION: Endemic.

Stag's tongue, ʻēkaha, *Elaphoglossum crassifolium*

DESCRIPTION: This is the most common stag's tongue fern in Hawaiʻi. Fronds are single fleshy blades with blunt tips and long, dark brown stalks. Blades reach over a foot (30 cm) in length with a stalk of almost equal length. Holding a frond up to the light reveals the network of veins that distinguishes it from other stag's tongue ferns, which have parallel veins. Often this species grows intermixed with other stag's tongue ferns. The scientific name means "thick-leaved stag's tongue."

HABITAT: This fern is found in the open and in shade from dry areas to wet forest, usually above 1,000 feet (305 m). It grows on the ground or as an *epiphyte*.

DISTRIBUTION: Endemic.

Hairy stag's tongue, *Elaphoglossum hirtum* subspecies *micans* (*Elaphoglossum paleaceum*)

DESCRIPTION: A fur of brown scales covers the fronds, making this fern easy to distinguish from other stag's tongue ferns. Fronds are supple, with branched veins that do not join together at the frond edge. The leafy section of the frond averages between 1/2 and 1 foot (15–30 cm) in length. The scientific name means "glittering, hairy stag's tongue."

HABITAT: This fern is common above 2,000 feet (610 m) up to 7,000 feet (2,135 m) elevation in both dry and wet forests, growing on the ground and perched on rocks and trees.

DISTRIBUTION: This subspecies is endemic.

SWORD FERNS
(Nephrolepidaceae)

Hawaiians named members of this family *kupukupu*, *'ōkupukupu*, and, on Ni'ihau, *ni'ani'au*. These names refer to any sword-shaped fern. The scientific name comes from the kidney ("nephro")-shaped *indusia* ("lepis") that form over the spores, opening toward the leaf edges. At least fifteen species grow in tropical regions and as far south as New Zealand and north to Japan. Sword ferns grow on trees, rocks, and the ground. *Kupu* means "to sprout" in Hawaiian. *Hula* altars were decorated with this fern so that knowledge would sprout. When adorning their wrists and ankles, *hula* dancers chant: "the fragrant *kupukupu* ferns of the heights of Kānehoa" (a hill on O'ahu named for Pele's father). Some botanists include sword ferns in the family *Dryopteridaceae*.

Sword fern, *kupukupu*, *'ōkupukupu*, *Nephrolepis cordifolia*

The species name means "heart-leaved."

DESCRIPTION: Fronds grow erect, reaching over 2 feet (61 cm) in length, and are narrower than those of other sword ferns, with more space between divisions. Fronds are glossy, yellowish green to dark green, and divided once. Ranks of kidney-shaped *indusia* form on either side of the *pinnae* midribs. Small, curly, brown scales form on the lower stalks.

HABITAT: This terrestrial and epiphytic fern grows in wet forest at elevations from 2,000 feet (610 m) to over 4,000 feet (1,220 m).

DISTRIBUTION: Indigenous.

Common sword fern, kupukupu, 'ōkupukupu, Nephrolepis exaltata

Some sources give it the Hawaiian name of *pāmoho*. The species name means "very tall."

DESCRIPTION: Fronds are yellowish green and long and narrow, resembling those of the cultivated Boston fern. Fronds are taller than those of *Nephrolepis cordifolia*, reaching 4 feet (1.2 m) in length. Its smooth surface distinguishes this native fern from a very common introduced sword fern, *Nephrolepis multiflora*, which has minute hairs on the upper surface of the fronds.

HABITAT: An epiphytic and terrestrial fern, *kupukupu* is found from sea level to elevations of 3,500 feet (1,065 m). It grows in sun or shade in dry to wet areas. Its introduced relative, *N. multiflora*, has been naturalized and is now very abundant.

DISTRIBUTION: Indigenous.

BLECHNUM FERNS
(Blechnaceae)

This widespread family takes its name from a Greek word for ferns, *blechnon*. Generally, blechnums are stiff and coarse, with salmon-colored young fronds. Their spore clusters form under protective flaps along lines parallel to the leaf veins, with each flap opening toward the midrib.

Of the 250 species in this family, about eight species grow in Hawai'i. Half of them belong to the genus *Sadleria*, which is unique to Hawai'i.

Blechnum occidentale

This fern is so widespread, even in remote parts of the Islands, that it might seem to be native, although it was introduced to Hawai'i from tropical South America around 1917. Its "western" origin led to its scientific name, *occidentale*.

DESCRIPTION: This fern's smooth, deep green fronds are divided once and grow to 18 inches (46 cm) long and 6 inches (15 cm) wide. The fronds end in a long point. Spores form in two parallel lines along the center of each division. The pink of young fronds helps to distinguish them from sword ferns, and the arrangement of spores is very different from that of its relative, *Doodia*.

HABITAT: This fern is common in disturbed places, especially banks along trails, at lower elevations and as high as 3,500 feet (1,065 m), usually in partial shade and damp conditions.

DISTRIBUTION: Naturalized.

Rasp fern, '*ōkupukupu lau'i'i*, *Doodia kunthiana*

Species of *Doodia* grow in humid, frost-free parts of Sri Lanka, New Zealand, Australia, and New Caledonia. Members of this genus have stiff, erect, small to medium-sized fronds. Oblong clusters

of spore capsules form one or more lines along the middle veins of each division. Samuel Doody, for whom this genus is named, was an English botanist and apothecary who lived until 1706. The species name honors Karl Sigmund Kunth, a botanist at the University of Berlin who lived from 1788 to 1850.

The Hawaiian name, *'ōkupukupu lau'i'i*, means "small-leaved *'ōkupukupu*." *'I'i* also means reddish brown and may refer to the color of the young fronds. The name *'ōkupukupu* refers to several narrow ferns that are divided once.

DESCRIPTION: Fronds grow up to 25 inches (64 cm) long and are deep green and leathery, unlike the softer fronds of sword ferns. Young fronds are often maroon in color. The shape of this once-divided fern resembles that of sword ferns (*Nephrolepis*). Unlike sword ferns, the upper surface of these ferns is bumpy above the *sori*. Oblong spore pockets line both sides of the midrib of each division. A protective flap *(indusium)* covers the *sori* and opens toward the division's midrib.

HABITAT: This is a terrestrial fern commonly found in lower to middle elevations, especially shaded dry and moist gullies.

DISTRIBUTION: Endemic.

'Ama'u, Sadleria cyatheoides

This genus was named for a professor of botany in Budapest, Joseph Sadler, who died in 1841. The species name describes the tree fern–like trunk of this blechnum fern. Only four other species are included in this genus, all of which are unique to Hawai'i. *'Ama'u* forms a trunk, and its smooth and leathery fronds are more finely divided than others in this family.

The Hawaiian name, ʻamaʻu, refers to most members of this endemic genus. The name ʻamaʻumaʻu refers to the young ferns. Young fronds are salmon-colored and turn dark, glossy green as they mature. These ferns were prevalent in many aspects of Hawaiian life and lore. Hawaiians mulched their dryland crops, fringed their thatched huts, and, during rituals, paved pathways for royalty with these tough fronds. Makers of bark cloth (kapa) glued pieces of cloth together with a slimy extract from these fronds, called palaholo. A red dye from the ʻamaʻu trunk was used to color kapa.

The mischievous Hawaiian demigod, Kamapuaʻa ("pig person"), took the form of this fern to escape from and entangle his enemies during his many escapades. Halemaʻumaʻu Crater in Hawaiʻi Volcanoes National Park, where Kamapuaʻa wooed the volcano goddess, Pele, is named "house of the ʻamaʻumaʻu" for the fern form of Pele's persistent suitor.

This hardy, common fern grows in Kamapuaʻa's wet mountain haunts. ʻAmaʻu forms a trunk up to 5 feet (1.5 m) tall, the heart of which Hawaiians ate in times of famine.

DESCRIPTION: Though similar to those of tree ferns, in ʻamaʻu the fronds are not as finely divided and are relatively narrower. They are glossy dark green above and whitish green below. A furry mass of golden pulu (hairlike flat scales) surrounds the bases of its dull green stalks. This pulu was once used for stuffing mattresses, as was tree fern pulu. Fronds can reach a length of 10 feet (3 m). Lines of spores run nearly the length of the leaflet segments.

HABITAT: A terrestrial fern, *'ama'u* grows in open or shady forests up to 2,000 feet (610 m) elevation. It grows to 7,000 feet (2,135 m) in subalpine scrub on Maui and Hawai'i and also on windward sea cliffs at less than 100 feet (30 m).

DISTRIBUTION: Endemic.

'Apu'u, Sadleria squarrosa

According to some sources, this fern is also called *'āpulu* in Hawaiian. The scientific name means "Sadler's [fern] with recurved edges."

DESCRIPTION: The fronds are leathery and green, except for a pink tinge when young. Fronds are divided twice, with small divisions whose edges curl under and are much shorter than those on *'ama'u* (*Sadleria cyatheoides*). *'Apu'u* grows 1 to 2 feet (30–61 cm) long, draped from steep faces. Golden brown scales cover the stalk of the fronds. Though not abundant, it can usually be found in its favored habitat.

HABITAT: *'Apu'u* grows on wet, shaded, decaying rock faces, often along streams, in high mountains above 2,000 feet (610 m).

DISTRIBUTION: Endemic.

WATER FERN
(Azollaceae)

This tiny fern prefers mild climates. Six species are recognized in the world, but only one grows in Hawai'i. It is also called pepperwort and mosquito fern. The scientific name, *Azolla*, means "kill (*olla*) when dry (*azo*)," referring to its aquatic habitat.

Water fern, *Azolla filiculoides*

DESCRIPTION: This is the smallest of ferns; its scalelike fronds are only 1/8 inch (3 mm) wide. It forms floating mats on taro paddies, ponds, and in slow-moving water. Each plant forms a tiny cluster of overlapping leaves. It seldom produces spores and usually reproduces by segments breaking off and developing into new plants. The species name means "fernlike."

HABITAT: This is an aquatic, lowland fern.

DISTRIBUTION: Introduced to the Islands in the 1930s, probably as a hitchhiker with other plants.

GLOSSARY

ECOSYSTEM   An interacting unit of organisms and their environment.

ENDEMIC   Organisms unique to one area, in this case the Hawaiian Islands or one particular island.

EPIPHYTIC   Growing on a tree, but not taking nutrients from its host.

FERN ALLY   Primitive relatives of ferns with scalelike leaves and with spores borne on spikes rather than leaves.

FROND   The leaf of a fern.

GAMETOPHYTE   Sometimes called a prothallus, this is an intermediate stage of a fern plant in which sexual reproduction takes place. Gametophytes are usually small and inconspicuous and grow in moist, shady places.

GENUS (plural GENERA)   A grouping of species with common traits, differing from other genera in a distinctive set of characteristics.

INDIGENOUS   Organisms naturally occurring in a given region (for example, Polynesia).

INDUSIUM (plural INDUSIA)   A protective flap that covers the cluster of spore capsules (sorus) on some ferns.

INTRODUCED   Brought to an area by humans, intentionally or accidentally (see naturalized).

KEIKI   A Hawaiian term for child or offspring, used to refer to new fern shoots that form on the fronds of ferns that reproduce asexually.

LEI   A Hawaiian garland of flowers, leaves, seeds, or fruit.

LOBE   A frond division that is not stalked or cut to the midrib.

NATURALIZED   An introduced plant that becomes established and reproduces in the wild without the aid of humans.

PENDENT   Hanging.

PINNA (plural PINNAE)   A division of the frond that is cut to the midrib and may be stalked. Fronds may be undivided, pinnate, bipinnate, or tripinnate.

PULU   Hairlike scales that form at the base of some fronds.

RHIZOME   A creeping stem from which fronds and roots grow. An upright stem, such as those found on tree ferns, is called a caudex.

RUPESTRAL   Growing on a rock.

SORUS (plural SORI)   A cluster of spore capsules (*sporangia*).

SPECIES   A grouping of organisms that share many characteristics and generally breed and reproduce only among their own kind (although hybrids sometimes do occur between certain related fern species).

SPORANGIUM (plural SPORANGIA)   The spore-bearing capsule, which is solitary or grouped in clusters to make up the *sori*. Most fern sporangia contain 64 spores.

SPORE   A single, microscopic reproductive cell that grows into a gametophyte. Fern spores are usually dispersed by air currents, sometimes for long distances.

STIPE   The supporting stalk of a frond.

STROBILUS (plural STROBILI)   A conelike spore-bearing body found on fern allies such as club mosses and *Selaginella* species.

SUBSPECIES (ssp.)   A division below the species level designating a population of individuals that are geographically or ecologically distinct from other populations of a species.

TERRESTRIAL   Growing on or from the ground.

VARIETY (var.)   A term used synonymously with subspecies, and alternatively as a lesser division of subspecies.

BIBLIOGRAPHY

Anderson, W. R., and M. R. Crosby. 1966. A revision of the Hawaiian species of *Elaphoglossum*. *Brittonia* 18: 380–397.

Bailey, L. H. 1963. *How Plants Get Their Names*. New York: Dover Publications.

Brownsey, P. J., and J. C. Smith-Dodsworth. 1989. *New Zealand Ferns and Allied Plants*. Auckland: David Bateman.

Degener, O. 1975. *Plants of Hawaii National Park Illustrative of Plants and Customs of the South Seas*. Ann Arbor: Braun-Brumfield.

Everett, T. H. 1981. *The New York Botanical Garden Illustrated Encyclopedia of Horticulture*. New York: Garland Publishing.

Fosberg, F. R. 1942. Uses of Hawaiian ferns. *American Fern Journal* 32(1): 15–23.

Grillos, S. J. 1966. *Ferns and Fern Allies of California*. Berkeley: University of California Press.

Lamoureux, Charles. 1988. Checklist of Hawaiian ferns. Unpublished manuscript.

MacDonald, M. A. 1978. *Ka Lei: The Leis of Hawaii*. Honolulu: Topgallant Publishing Co.

May, L. W. 1978. The economic uses and associated folklore of ferns and fern allies. *The Botanical Review* 44(4): 491–528.

Medeiros, A. C., et al. 1992. Notes of the status of an invasive Australian tree fern (*Cyathea cooperi*) in Hawaiian rain forests. *American Fern Journal* 82(1): 27–33.

Olson, W. W. 1977. *The Fern Dictionary*. Los Angeles: Los Angeles International Fern Society.

Parsons, F. T. 1961. *How to Know the Ferns*. New York: Dover Publications.

Plowden, C. C. 1970. *A Manual of Plant Names*. New York: Philosophical Library.

Stafleu, F. A., and R. S. Cowan. 1986. *Taxonomic Literature*. The Netherlands: Frans A. Stafleu.

Wagner, W. H. 1950. Ferns naturalized in Hawaii. Occasional Papers of the Bernice Pauahi Bishop Museum 20: 95–121.

———. 1987. Status of the Hawaiian fern flora. *Fiddlehead Forum* 15: 11–14.

Wilson, W. F. (ed.). 1922. *Extracts from the Diary of James Macrae*. Honolulu: privately published.

INDEX

adder's tongue, 8, 25–26
Adenophorus: hymenophylloides, 29; *pinnatifidus*, 30; *tamariscinus*, 31; *tamariscinus* var. *tamariscinus*, 31
Adiantaceae, 41
Adiantum: capillus-veneris, 41; *cunaetum*, 42; *hispidulum*, 44; *raddianum*, 42
ʻae, 36–37
ʻāhina kuahiwi, 68
ʻakole, 69
ʻākōlea, 64–65
ʻamaʻu, 4, 5, 8, 10, 78–80; ʻamaʻumaʻu, 79
ʻanaliʻi, 61
ʻāpulu, 80
ʻapuʻu, 80
Aspidiaceae, 68
Aspleniaceae, 60
Asplenium: adiantum-nigrum, 60–61; *lobulatum*, 6, 61; *nidus*, 35, 62; *normale*, 6, 62–63; X *adiantum-nigrum*, 60
Athyriaceae, 63
Athyrium: japonicum, 63–64; *microphyllum*, 64–65; *sandwichianum*, 66
Azolla filiculoides, 3, 81
Azollaceae, 80

bamboo fern, 45
bird's nest fern, 35, 62
Blechnaceae, 76
Blechnum ferns, 76
Blechnum occidentale, 77
Boston fern, 76
bracken ferns, 10, 54–55

Cheilanthes decipiens, 37
Christella: cyatheoides, 57–58, 59; *dentata*, 58; *parasitica*, 58
Christmas tree club moss, 19–20
Cibotium: glaucum, 53; *menziesii*, 52–53
cliff brake, 38
club mosses, 6, 19–20
Coniogramme pilosa, 45

Cretan brake, 39
Cyatheaceae, 51
Cyathea sp., 57, 78; *cooperi*, 4
Cyclosorus interruptus, 58–59
Cyrtomium: caryotideum, 68–69; *falcatum*, 68

Dennstaedtiaceae, 53–54
Deparia: petersenii, 63; *prolifera* var. *marginale*, 6, 66
Dicksoniaceae, 51
Dicranopteris linearis, 26, 28
Diplazium sandwichianum, 66–67
Diplopterygium pinnatum, 28
Doodia kunthiana, 77–78
Doryopteris decipiens, 37–38
Dryopteridaceae, 63, 68, 71, 74
Dryopteris: fusco-atra, 69; *glabra*, 67; *unidentata*, 69; *wallichiana*, 69

ecosystem, 4, 83
ʻēkaha, 35, 62, 71, 73; ʻēkaha ʻākōlea, 35
ʻēkahakaha, 62
Elaphoglossaceae, 71
Elaphoglossum: aemulum, 71; *alatum*, 15; *crassifolium*, 73; *hirtum* ssp. *micans*, 14, 73–74; *paleaceum*, 73
endemic, 3, 83
epiphytic, 8, 83

false staghorn fern, 15, 26–29
fern allies, 3, 6, 19–22, 83
filmy ferns, 47–51
finger ferns, 29–32
five-finger maidenhair, 44
frond, 6, 8, 10, 83

gametophyte, 6, 83
genus, 14–15, 83
giant *uluhe*, 28
Gleicheniaceae, 15, 26
golden fern, 46–47

Index

Gonocormus minutus, 47–48
Grammitaceae, 29
Grammitis tenella, 31–32

hairy stag's tongue, 73–74
hanging adder's tongue, 25–26
hāpu'u, 5, 51–53; hāpu'u 'i'i, 52–53
hare's foot fern, 33–34
Hawaiian bow fern, 68–69
hay-scented ferns, 53–54
Hemionitidaceae, 45
hō'i'o, 66–67, 69
hula, 5, 41, 54, 56, 76
huluhulua'iole, 19
huna palai, 29
Huperzia phyllantha, 20
Hymenophyllaceae, 47
Hypolepidaceae, 54

'i'i, 52
indigenous, 3, 83
indusium, 10, 83
introduced, 3–4, 83
'iwa'iwa, 14, 37–38, 41–44, 60–61
'iwa'iwa launui, 70

kā'ape'ape, 68–69
kalamoho, 38
keiki, 6, 61, 83
kihe, 32
kikawaiō, 4, 57
kīlau, 55; kīlau pueo, 55
kolokolo, 31
kumuniu, 37
kupukupu, 74, 76; kupukupu makali'i, 57

lace ferns, 56–57
lady ferns, 63–67
Lamoureux, Charles, 15
laua'e, 5, 34–35; laua'e haole, 33–34
laukahi, 4, 46
laukahi nunui, 71
lei, 5, 22, 34–35, 83
Lellingeria saffordii, 32
lepelepeamoa, 22
Lepisorus thunbergianus, 35
Lindsaeaceae, 56
lobe, 8, 83
lo'ulu, 45
Lycopodiaceae, 19

mahinalua, 31–32
maiden ferns, 57–60
maidenhair ferns, 41–44
maile-scented fern, 34–35
manawahua, 37
Mecodium recurvum, 48–49
Microlepia strigosa, 54
Microsorum spectrum, 33
moa, 4, 21, 25
moa nahele, 21
mosquito fern, 80

naturalized, 4, 83
neke, 58–59
Nephrolepidaceae, 74
Nephrolepis: cordifolia, 74, 76; *exaltata*, 12, 76; *multiflora*, 76
ni'ani'au, 74

Odontosoria chinensis, 56
'ōhi'a kū, 48–49
'ōkupukupu, 12, 74–78; 'ōkupukupu lau'i'i, 77–78
Oligadenus pinnatifidus, 30
'opeha, 71
Opihioderma pendula, 25
Ophioglossaceae, 25
Ophioglossum pendulum spp. *falcatum*, 4, 15, 25
'ōwali, 39

pai, 29–30
pai'ā, 55
pai'i'iha, 58
pākahakaha, 35–36
pala'ā, 5, 56–57
palaholo, 79
palai, 30, 49, 54; palai hinahina, 49; palai lā'au, 30; palai'ula, 54
palapalai, 5, 46, 54
Palhinhaea cernua, 19
pe'ahi, 33
Pellaea ternifolia, 38
pendent, 83
pepperwort, 80
Phlebodium aureum, 33–34
Phlegmariurus phyllanthus, 20
Phymatosorus scolopendria, 34–35
pi'ipi'ilaumanamana, 61
pinna, 8, 83

pipi, 21
Pityrogramma: *austroamericana*, 47; *calomelanos*, 46–47
Pleopeltis thunbergiana, 35–36
Pneumatopteris sandwicensis, 57, 59–60
pohole, 66
Polypodiaceae, 33
Polypodium pellucidum, 36–37
polypody ferns, 33–37
Pseudophegopteris keraudreniana, 39
Psilotaceae, 21
Pteridaceae, 37, 39, 41, 45
Pteridium: *aquilinum*, 55; *aquilinum* ssp. *decompositum*, 55; *decompositum*, 55
Pteris: *cretica*, 10–12, 39; *excelsa*, 39–40; ferns, 39–40; *irregularis*, 39; X *hillebrandii*, 39
pua'akuhinia, 35
puapua moa, 15, 25–26
pulu, 5, 51, 53, 79

rasp fern, 77–78
rhizome, 12, 84
rupestral, 84

Sadleria: *cyatheoides*, 78–80; *squarrosa*, 80
Selaginella arbuscula, 22
Selaginellaceae, 22
silver fern, 46–47
Sinopteridaceae, 37
sorus (sori), 6, 10, 12, 84
species, 3, 12, 14–15, 84
Sphaerocionium lanceolatum, 49
Sphenomeris chinensis, 56
spikemoss, 22; branched, 22
spleenwort ferns, 60–63
sporangium (sporangia), 6, 84
spore, 3–4, 6, 10, 12, 19, 84

stag's tongue ferns, 71–74; hairy, 73–74
Stegnogramma stegnogrammoides, 59
Sticherus owyhensis, 15, 28–29
stipe, 84
strobilus, 20, 84
subspecies, 14, 84
swamp cyclosorus, 58–59
sword ferns, 12, 74–76; common, 76

Tectaria: *cicutaria* ssp. *gaudichaudii*, 70; *gaudichaudii*, 70–71
terrestrial, 8, 84
Thelypteridaceae, 57
Thelyptris: *cyatheoides*, 57; *dentata*, 58; *interruptus*, 58–59;
tree ferns, 4–5, 8, 16, 51–53; Australian, 4; Hawaiian, 52–3

uluehe, 26
uluhe, 3–4, 8, 26–29; giant, 28; *uluhe lau nui*, 28
unuhe, 26

Vandenboschia: *cyrtotheca*, 50–51; *davallioides*, 50–51
variety, 14, 84
Venus hair, 41–42

Wagner, Warren H. and Florence, 15
wahine noho mauna, 31
waimakanui, 39–40
water fern, 3, 80–81
wawae'iole, 4, 19–20
whisk fern, 21
wood ferns, 68

Xiphopteris saffordii, 32